The Nalco Guide
to Boiler
Failure Analysis

Nalco Chemical Company

Authored by

Robert D. Port

Harvey M. Herro

McGraw-Hill, Inc.

New York   St. Louis   San Francisco   Auckland   Bogotá
Caracas   Lisbon   London   Madrid   Mexico   Milan
Montreal   New Delhi   Paris   San Juan   São Paulo
Singapore   Sydney   Tokyo   Toronto

Library of Congress Cataloging-in-Publication Data

Port, Robert D.
   The Nalco guide to boiler failure analysis / the Nalco Chemical
Company; authored by Robert D. Port, Harvey M. Herro.
      p.   cm.
   Includes index.
   ISBN 0-07-045873-1
   1. Steam-boilers — Failures — Handbooks, manuals, etc.   I. Herro,
Harvey M.   II. Nalco Chemical Company.   III. Title.
TJ298.P67   1991
621.1′83 — dc20   90-40834
                                                                        CIP

Dedicated to Dr. William K. Baer, a Motivator

   89 KPKP 99

ISBN 0-07-045873-1

*The sponsoring editor for this book was Robert W. Hauserman, the editing
supervisor was Alfred Bernardi, the designer was Naomi Auerbach, and
the production supervisor was Thomas G. Kowalczyk. It was set in Century
Schoolbook by Progressive Typographers.*

Printed and bound by Kingsport Press.

Contents

Preface

This book is a comprehensive, authoritative field guide to boiler failures. Contained is information gathered during forty years of failure analysis at Nalco Chemical Company involving over 10,000 case histories. Failures in boilers of virtually all pressures and construction are represented. Failures in nearly all boiler locations, as well as some preboiler and afterboiler regions, are discussed. Many photographs of actual failures illustrate the text.

Critical, unprejudiced observation, bolstered by experience and common sense, are the most important factors in the correct identification of a boiler failure. This book alone cannot substitute for the trained, experienced analyst. Failure analysis is a learned skill. This book will help the novice and more experienced observer select the most likely failure modes quickly and simply. By direct comparison with the many illustrations and photos of real failures, the user will learn to identify many problems and pinpoint likely failure locations. Explanation of failures will be easier, and the interpretation and discussion of failure-analysis reports will be facilitated.

The book is not intended to be a substitute for rigorous failure analysis. Rather, the information should serve as an aid to in-plant investigation and as a readily accessible reference source in the field. Although the skilled observer can detect and correctly diagnose many failures based solely upon visual inspection, often there is no substitute for laboratory investigations.

Topics include overheating, chemical corrosion on both the fire side and water side, cracking, deposition, erosion, and manufacturing defects. Most failure modes or topics are discussed in seven sections including:

Locations: Failure locations are pinpointed.

General description: Failure mechanisms are described. The causes and sources of attack are discussed.

Critical factors: The most important or especially critical factors influencing failure are listed. Attack morphology and identification are stressed.

Identification: Visual inspection procedures are emphasized. The methods by which detailed metallographic procedures might elucidate failure mechanisms are also discussed.

Elimination: Necessary preventive and/or corrective steps to eliminate or reduce damage are listed and explained.

Cautions: Comparisons are drawn between failure mechanisms that produce damage with a similar appearance.

Related problems: Related problems and damages are described.

Case histories illustrate each chapter. Examples are chosen to illustrate both similarities and differences between actual and "classical" failures. These histories provide diverse perspectives on how failures actually occur.

Nalco Chemical Company is currently preparing a failure-analysis reference book dealing with cooling-water failures, which will follow the same format as this text. This second book should be available in 1991.

Acknowledgments

Thanks are extended to Nalco Chemical Company for its generous support during this work. Our colleague, James J. Dillon, is the recipient of special appreciation for his contribution of selected photos and case histories. Special thanks is due Michael J. Danko for several photos. Warmest thanks are given to Ted R. Newman, our mentor. Credit is due David L. Pytynia for his efforts during project initiation and Donald G. Wiltsey for his excellent review of the manuscript. Valerie J. Vujtech provided invaluable assistance in manuscript preparation. Acknowledgment also goes to Pamela J. Entrikin for final manuscript edits, cover design direction, and her assistance in coordinating production of this book with McGraw-Hill.

Water-Formed
and Steam-Formed
Deposits

Locations

Deposits can occur anywhere water or steam is present in a boiler. While wall and screen tubes are usually the most heavily fouled, roof and floor tubes often contain deposits as well. Superheaters and reheaters frequently contain deposits that are formed elsewhere and are carried into the systems with boiler water. Steam is not often generated in economizers. Deposits in these tubes are usually made up of corrosion products moved from their origination sites.

Deposition can be substantial during steam generation. Tube orientation can influence the location and amount of deposition. Deposits are usually heaviest on the hot side of steam-generating tubes. Because of steam channeling, accumulations are often heavier on top portions of horizontal and slanted tubes. Also, deposition often occurs immediately downstream from circumferential-weld backing rings, which disturb flow and are favored sites for steam blanketing. Because deposits tend to concentrate in the hottest regions of steam-generator tubes, those tubes near the bottom rear wall of boilers using chain-grate stokers, and screen tubes are susceptible to deposition. Coarse particulate matter is likely to be found in horizontal runs and where flow velocity is small.

Most economizers are designed to operate without producing steam. Waterborne deposits usually enter the economizer from sources such as the returned condensate (usually not polished). Oxides formed as a result of elevated oxygen concentrations prior to or inside the economizer may be moved and deposited in the economizer.

Mud and steam drums often contain deposits. Because drums are readily accessible, a visual inspection can provide many details about water chemistry and deposition processes. For example, sparkling black magnetite crystals may precipitate in steam drums when iron is released by the decomposition of organic complexing agents.

Superheater deposits are caused by carryover of boiler water, sometimes associated with foaming or high water levels. Such deposits will usually be concentrated near the superheater inlet or in nearby pendant U-bends. Contaminated attemperation water can also add deposits immediately downstream from the introduction point. Chip scale and exfoliated oxide particles can be blown through the superheater, accumulating in pendant U-bends, or even more seriously, can be carried into turbines.

General Description

The term *deposits* refers to materials that originate elsewhere and are conveyed to a deposition site. Deposits cannot be defined as corrosion products formed in place, although corrosion products formed elsewhere and then deposited do qualify. Oxides formed from boiler metal are not deposits unless they have been moved from their origination sites. *This distinction is fundamental.*

Boiler deposits come from four sources: water borne minerals, treatment chemicals, corrosion products (preboiler and boiler), and contaminants. Deposits from these sources may interact to increase deposition rates, to produce a more tenacious layer, and to serve as nucleation sites for deposit formation. Such species include (but are not limited to) metal oxides, copper, phosphates, carbonates, silicates, sulfates, and contaminants, as well as a variety of organic and inorganic compounds.

One deposition process involves the concentration of soluble and insoluble species in a thin film bordering the metal surface during steam-bubble formation (Fig. 1.1). Material segregates at the steam/water interface, moves along the interface, and is deposited at the bubble base as the bubble grows. Other deposit mechanisms involve precipitation from solution and settling of large particulate matter. Inverse-temperature solubility leads to deposition where heat transfer is great.

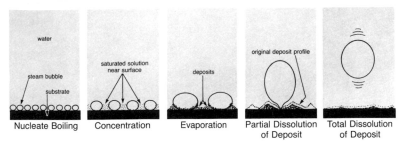

Nucleate Boiling | Concentration | Evaporation | Partial Dissolution of Deposit | Total Dissolution of Deposit

Figure 1.1 Five instants in the life of a steam bubble.

The tendency to form deposits is related to localized heat input, water turbulence, and water composition at or near the tube wall. When a steam bubble becomes dislodged from a tube wall, the deposits are washed with water. The rate at which the deposit builds depends on the rate of bubble formation and the effective solubility of the deposit. In cases of high heat input, a stable steam blanket can form and cause concentration of water-soluble material (Fig. 1.2). Steam-blanket deposits do not redissolve, because the surface cannot be washed while blanketed with steam (Fig. 1.3). Steam blanketing also results from surface irregularities, which disturb water flow. Downstream of such irregularities, low-pressure areas are formed, favoring steam buildup and consequently deposit formation.

Steaming does not often occur in economizers. Deposits are usually iron-rich. The iron oxide is produced in the preboiler system or is formed in the economizer itself (Figs. 1.4 and 1.5).

Thermal stresses aid oxide spalling. Exfoliation of thermally formed oxides in superheaters and reheaters can cause accumulation of oxide

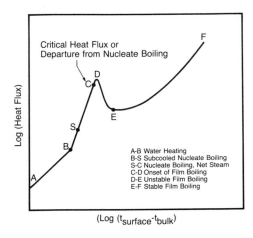

Critical Heat Flux or Departure from Nucleate Boiling

A-B Water Heating
B-S Subcooled Nucleate Boiling
S-C Nucleate Boiling, Net Steam
C-D Onset of Film Boiling
D-E Unstable Film Boiling
E-F Stable Film Boiling

Log (Heat Flux)

(Log ($t_{surface}$-t_{bulk}))

Figure 1.2 Heat transfer to water and steam in a heated flow channel. Relation of heat flux to temperature difference between channel-wall and bulk-water or steam temperature. (*Courtesy Babcock and Wilcox Company*, Steam/Its Generation and Use, *New York, 1972*).

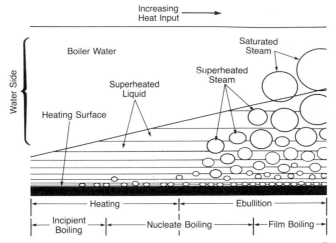

Figure 1.3 Transition from heating to boiling (ebullition) as wall temperature increases.

chips in U-bends and long horizontal runs (Figs. 1.6 and 1.7). Scaling temperatures (the temperatures above which oxide formation is rapid) for a variety of alloys are shown in Table 2.1. Waterborne deposits can also accumulate because of carryover due to foaming, "gulping," and water washing. Steam-soluble species can be carried through superheaters and

Figure 1.4 Thick layer of friable iron oxide in an economizer tube. Most iron was carried into the economizer from the preboiler system.

Figure 1.5 Fragmented hematite layer on an internal surface of an economizer. Most oxide was formed in place, but some was moved as fragments and chips.

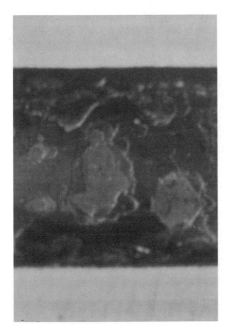

Figure 1.6 Exfoliated magnetite patches in a tube from the primary superheater of a utility boiler. Magnetite chips can be carried into turbines and cause severe damage (see Fig. 1.7).

Figure 1.7 Fine-mesh turbine inlet screen. Pieces of exfoliated oxide from superheater tubing are wedged into screen openings. (Magnification: 7.5×.)

deposited on turbines. If chlorides and sulfates are present, hydration can cause severe corrosion due to hydrolysis (Figs. 1.8 and 1.9).

Critical Factors

The rate at which deposits form on heat-transfer surfaces is controlled mainly by the solubility and physical tenacity of the deposit and the amount of water washing that occurs where steam is generated. Solubility,

Figure 1.8 Severely corroded blades from high-pressure condensing stage of a turbine. Deposits were removed to reveal attack.

Figure 1.9 Deposits on turbine blades before cleaning, as in Fig. 1.8. (Magnification: 7.5×.)

tenacity, and water washing, in turn, depend on other factors such as dissolved-solids concentration, formation temperature, agglomeration morphology, and turbulence. However, the prerequisite factor for significant deposit formation is usually steam formation. In fact, deposits can form even when steaming is slight (Fig. 1.10). As long as nucleate boiling is

Figure 1.10 Tube sections virtually plugged with deposits. The tube on the right is from a low-pressure boiler and is fouled with almost pure calcium carbonate. The center tube contains silicates, phosphates, and other compounds; fouling occurred on standby service. The section on the left is reddened by almost 20% elemental copper. Such heavy deposition can occur only when internal pressures are low. Otherwise rupture would occur.

occurring, heat transfer is controlled by tube-wall and deposit thermal conductivities and by the gas-side temperature. The thermal conductivities of several boiler deposits and alloys are given in Table 1.1.

Salts having inverse-temperature solubility deposit readily on heat-transfer surfaces. For example, calcium sulfate and calcium phosphate deposit preferentially in hot areas as temperatures increase. Eventually steam blanketing occurs and evaporation to dryness causes concentration of species having normal-temperature solubility (Fig. 1.11). Frequently, the deposits of the most insoluble materials are found in water-cooled tubes having the highest heat transfer, such as screen tubes. When evaporation to dryness occurs, both soluble and insoluble deposits are usually found together.

TABLE 1.1   Thermal Conductivities (ϵ) of Alloys and Deposits

	Thermal conductivity	
	$W/(m^2 \cdot °C)$	$(Btu \cdot ft.)/(h \cdot ft^2 \cdot °F)$
Alloys		
304 Stainless	22(500°C), 16(1000°)	12.5(932°F), 9.4(212°F)
410 Stainless	28(400°C), 25(1000°C)	16.1(752°F), 14.4(212°F)
Alloy steel (0.34 C, 0.55 Mn, 0.78 Cr, 3.53 Ni, 0.39 Mo, 0.05 Cu)	36*	21*
Copper	420*	240*
Carbon steel (0.23 C, 0.64 Mn)	55*	32*
Aluminum	235*	136*
Deposits		
Aluminum oxide, fused (Al_2O_3)	3.60	2.1
Analcite ($Na_2O \cdot Al_2O_3 \cdot 4SiO_3 \cdot 2H_2O$)	0.19	0.76
Calcium carbonate ($CaCo_3$)	0.14	0.56
Calcium phosphate [$Ca_3(PO_4)_2$]	0.55	2.20
Calcium sulfate ($CaSO_4$)	0.21	0.83
Ferric oxide (Fe_2O_3)	0.09	0.35
Magnesium oxide (MgO)	0.17	0.69
Magnesium phosphate [$Mg_3(PO_4)_2$]	0.33	1.30
Magnetite (Fe_3O_4)	0.45	1.8
Porous materials	0.01	0.06
Quartz (SiO_2)	0.24	0.97
Serpentine ($3MgO \cdot 2SiO_2 \cdot 2H_2O$)	0.16	0.63

* Room temperature.

NOTE: $(Btu \cdot ft.)/(h \cdot ft^2 \cdot °F) = 1.73 \ W/(m^2 \cdot °C)$

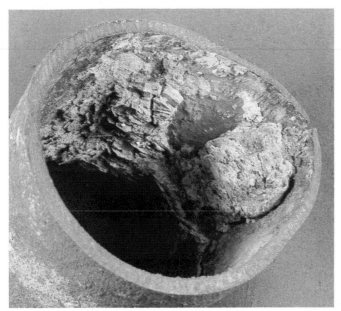

Figure 1.11 Sodium hydroxide–rich deposit on fire side of wall tube from 600-psi boiler. Note the bulge due to long-term overheating.

Even a relatively small amount of deposit can cause wall temperatures to rise considerably. As wall temperature rises, the tendency to steam-blanket increases (Fig. 1.3). Blanketing decreases heat flow, potentially causing overheating and rupture (see Chap. 2, "Long-Term Overheating").

Water quality also has a significant influence on deposition. Suggested acceptable feedwater quality as a function of boiler pressure is shown in Table 1.2. This table indicates that fewer contaminants can be tolerated at high boiler pressures. The insulating effects of deposits become less tolerable as pressures rise, because overheating is more likely. Maximum acceptable concentrations of boiler-water salines are shown in Table 1.3. The recommended dissolved solids decrease by a factor of 100 when pressure rises from 100 psi (0.69 MPa) to 2000 psi (13.8 MPa). Allowable silica levels decrease by a factor of 250 and suspended solids by a factor of 500.

A rule of thumb concerning tube cleanliness suggests that high-pressure boilers (pressures greater than 1800 psi or 12.4 MPa) are considered relatively clean if less than 15 mg/cm^2 ($\sim$14 g/ft^2) of deposits are present on water-cooled tubes. This amount of deposit is typical of almost all kinds of clean boilers, regardless of water chemistry, boiler type, or fuel. In some boilers, porous magnetite layers up to 11 mg/cm^2 (10 g/ft^2) produce no significant impairment of heat transfer. Moderately dirty boiler tubes

TABLE 1.2   Recommended Feedwater Quality

Pressure (psi)	Silica range (ppm)	Total hardness	Maximum (ppm)		
			Oxygen	Iron*	Copper*
100	15–25	75.00	——	——	——
200	10–20	20.00	——	——	——
300	7.5–15	2.00	——	——	——
500	2.5–5.0	2.00	0.030	——	——
600	1.3–2.5	0.20	0.030	——	——
750	1.3–2.5	0.10	0.030	0.050	0.020
900	0.8–1.5	0.05	0.007	0.020	0.015
1000	0.2–0.3	0.05	0.007	0.020	0.015
1500	0.3 max	0.00	0.005	0.010	0.010
2000	0.1 max	0.00	0.005	0.010	0.010
2500	0.05 max	0.00	0.003	0.003	0.002
3200+	0.02 max	0.00	0.002	0.002	0.001

* In modern industrial boilers, which have extremely high rates of heat transfer, these concentrations should be essentially zero. Similarly, total hardness should not exceed 0.3 ppm $CaCO_3$, even at the lower pressures; suspended solids in the feedwater should be zero if possible.

SOURCE: Courtesy of Chemical Publishing Company, *The Chemical Treatment of Boiler Water,* James W. McCoy, New York,1981.

NOTE: 1 psi = 0.006895 MPa

contain 15 to 40 mg/cm^2 (14 to 37 g/ft^2) of deposits, and boiler tubes containing more than 40 mg/cm^2 (37 g/ft^2) are considered very dirty. Guidelines for high-pressure boilers are listed in Table 1.4. Heat transfer is severely reduced when deposit loading becomes excessive. One large boiler manufacturer recommends boiler cleaning at 32 mg/cm^2 (30 g/ft^2) for lower-pressure boilers. Prolonged operation above the maximum deposit loadings may produce serious corrosion and overheating failures. However, the weight of deposits alone does not always accurately indicate the tendency to overheat. Deposit composition and morphology also influence heat transfer.

Identification

Boiler treatment chemicals, feedwater composition, and heat input affect deposition. At pressures of 2500 psi (17.2 MPa) or more, hydrazine is normally used. Demineralization and condensate polishing are other commonly used water-treatment practices. This means deposits are likely to contain only iron oxide produced by the corrosion of internal surfaces, and possibly copper, nickel, or other metals and contaminants (Fig. 1.12); no water-treatment chemicals or their reaction products will normally be found. At lower pressures, a variety of species can occur. Typical com-

TABLE 1.3 Recommended Concentration of Boiler Salines

Pressure (psi)	Saturation temperature (°F)	Maximum (ppm)				Sludge conditioners		Residual phosphate	Range (ppm)	Residual hydrazine
		Dissolved solids	Suspended solids*	Total alkalinity†	Silica	Natural	Synthetic		Residual sulfite	
100	328	5000.00	500	900	250.00	150	15	NR‡	90–100	NR
200	382	4000.00	350	800	200.00	150	15	40–50	80–90	NR
300	417	3500.00	300	700	175.00	100	15	30–40	60–70	NR
500	467	3000.00	60	600	40.00	70	15	25–30	45–60	NR
600	486	2500.00	50	500	35.00	70	10	20–25	30–45	NR
750	510	2000.00	40	300	30.00	NR	10	15–20	25–30	NR
900	532	1000.00	20	200	20.00	NR	5	10–15	15–20	0.10–0.15
1000	545	500.00	10	50	10.00	NR	3	5–10	NR	0.10–0.15
1500	596	150.00	3	0	3.00	NR	NR	3–6	NR	0.05–0.10
2000	636	50.00	1	0	1.00	NR	NR	1–3	NR	0.05–0.10
2500	668	10.00	0	0	0.50	NR	NR	NR	NR	0.02–0.03
3200‡	705	0.02	0		0.02	NR	NR	NR	NR	0.01–0.02

* Guidelines for pressures from 100 to 900 psi apply to conventional field-erected boilers with moderate rates of heat transfer, say 50,000 Btu/(ft² · h). At high rates characteristic of packaged boilers, large amounts of insoluble material cannot be managed effectively by a dispersant presently available.

† Zero alkalinity refers to hydroxide ion, i.e., Pa and M alkalinites (2P-M). There is always some alkalinity produced by ammonia, hydrazine, morpholine, or other bases.

‡ NR = not recommended.

SOURCE: Chemical Publishing Company, *The Chemical Treatment of Boiler Water*, James W. McCoy, New York, 1981.

TABLE 1.4   Boiler-Tube Cleanliness

| | Deposits (mg/cm²) | | |
Boiler type	Clean	Moderately dirty	Very dirty
Supercritical units	<15	15–25	>25
Subcritical units (1800 psi and higher)	<15	15–40	>40

NOTE: 1 mg/cm² ~ 1 g/ft².
    1 psi = 0.006895 MPa
SOURCE: K. L. Atwood, and G. L. Hale, *A Method for Determining Need for Chemical Cleaning of High-Pressure Boilers,* presented to American Power Conference, Chicago, Illinois, April 20–22, 1971.

pounds, and their likely locations and deposit characteristics can be found in Table 1.5.

Iron oxides

A smooth, black, tenacious, dense magnetite layer naturally grows on steel under reducing conditions found on boiler water-side surfaces (Fig. 1.13). Magnetite forms by direct reaction of water with the tube metal. In higher-pressure boilers, the magnetite contains two layers, which ordinarily can be seen only by microscopic examination. Particulate iron oxide can deposit on top of the smooth, thermally formed magnetite layer if settling rates are high and/or if steaming is appreciable (Fig. 1.14). Usually, coarse, particulate magnetite will not tenaciously adhere to surfaces unless intermixed with other deposits. Sparkling, black, highly crystalline magnetite needles will often be present near caustic corrosion sites. Similar magnetite crystals can sometimes produce a sparkling surface coating on steam

Figure 1.12 Elemental copper on wall tube from a high-pressure utility boiler.

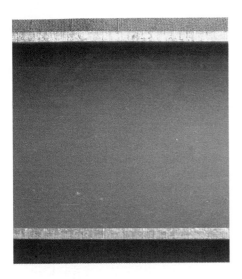

Figure 1.13 Smooth, tenacious, black, magnetite layer on a well-protected boiler tube.

drums and tubes (Fig. 1.15 and 1.16). Scraping small amounts of this material from surfaces and exposing these particles to a magnet usually indicates whether iron is present.

Hematite formation, on the other hand, is favored at somewhat lower temperatures and higher oxygen concentrations. Hematite is a binder species that tends to accumulate and hold other materials in the deposit. Hematite can be red if formed where oxygen concentrations are high. Hematite usually is present in economizers in conjunction with oxygen corrosion (Fig. 1.5).

Other metals and their oxides

Copper is deposited either by direct exchange with iron or by reduction of copper oxide by hydrogen evolved during corrosion. It is common to see large, reddish stains of elemental copper intermixed with corrosion products such as magnetite and hematite near caustic corrosion sites because of the hydrogen generation associated with attack (Fig. 1.12). The reddish color superficially resembles hematite. Elemental copper can be easily discriminated from other material by a silver nitrate test. A single drop of silver nitrate will precipitate white silver crystals almost immediately if elemental copper is present. Copper oxide formed under boiler conditions is black and nonmagnetic. Galvanic corrosion associated with copper deposits (either elemental copper or oxide) is exceedingly rare in well-passivated boilers.

Zinc and nickel oxides will sometimes be found in conjunction with copper deposits. This is to be expected, since zinc and nickel are often

TABLE 1.5 Components of Water-Formed Deposits

Mineral	Formula	Nature of deposit	Usual location and form
Acmite	$Na_2O \cdot Fe_2O_3 \cdot 4SiO_2$	Hard, adherent	Tube scale under hydroxyapatite or serpentine
Alpha quartz	SiO_2	Hard, adherent	Turbine blades, mud drum, tube scale
Amphibole	$MgO \cdot SiO_2$	Adherent binder	Tube scale and sludge
Analcite	$Na_2O \cdot Al_2O_3 \cdot 4SiO_2 \cdot 2H_2O$	Hard, adherent	Tube scale under hydroxyapatite or serpentine
Anhydrite	$CaSO_4$	Hard, adherent	Tube scale, generating tubes
Aragonite	$CaCO_3$	Hard, adherent	Tube scale, feed lines, sludge
Brucite	$Mg(OH)_2$	Flocculant	Sludge in mud drum and waterwall headers
Copper	Cu	Electroplated layer	Boiler tubes and turbine blades
Cuprite	Cu_2O	Adherent layer	Turbine blades, boiler deposits
Gypsum	$CaSO_4 \cdot 2H_2O$	Hard, adherent	Tube scale, generating tubes
Hematite	Fe_2O_3	Binder	Throughout boiler
Hydroxyapatite	$Ca_{10}(OH)_2(PO_4)_6$	Flocculant	Mud drum, waterwalls, sludge
Magnesium phosphate	$Mg_3(PO_4)_6$	Adherent binder	Tubes, mud drum, waterwalls
Magnetite	Fe_3O_4	Protective film	All internal surfaces
Noselite	$3Na_2O \cdot 3Al_2O_3 \cdot 6SiO_2 \cdot Na_2SO_4$	Hard, adherent	Tube scale
Pectolite	$Na_2O \cdot 4CaO \cdot 6SiO_2 \cdot H_2O$	Hard, adherent	Tube scale
Serpentine	$3MgO \cdot 2SiO_2 \cdot H_2O$	Flocculant	Sludge
Sodalite	$3Na_2O \cdot 3Al_2O_3 \cdot 6SiO_2 \cdot 2NaCl$	Hard, adherent	Tube scale
Xonotlite	$5CaO \cdot 5SiO_2 \cdot H_2O$	Hard, adherent	Tube scale

SOURCE: Courtesy of Chemical Publishing Company, *The Chemical Treatment of Boiler Water*, James W. McCoy, New York, 1981.

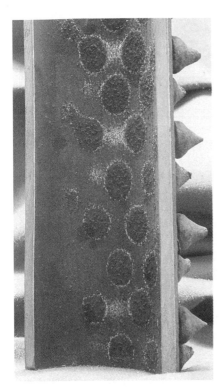

Figure 1.14 Spots of particulate iron oxide deposit mirroring stud locations on a wall tube of a recovery boiler. The studs, if exposed, concentrate heat, causing higher rates of steam generation and deposition. Note the white deposits of sodium hydroxide encircling each powdery magnetite spot.

Figure 1.15 Sparkling black magnetite crystals on internal surface of a wall tube. Crystals are formed by precipitation on surfaces, or are transported from preboiler areas.

Figure 1.16 Patch of sparkling magnetite crystals associated with caustic corrosion site. Such crystals are often found near sites of acid and caustic corrosion.

present in brass and cupronickels used in condensers and feedwater heaters. Usually, these elements are deposited at lower concentrations than copper. Nickel oxide has been indicated as a "binder" compound promoting tenacious deposition. However, nickel is usually present in relatively low concentrations and can be detected by chemical analysis using x-ray fluorescence or diffraction.

Salts

The least soluble compounds deposit first when boiling occurs. Calcium carbonate deposits quickly, forming a white, friable deposit that will effervesce when exposed to hydrochloric acid (Fig. 1.17). Calcium sulfate requires a higher degree of concentration to deposit than carbonate. The presence of phosphates can sometimes be inferred by friable (but less so than calcium carbonate) deposits.

Magnesium phosphate is a binder that can produce very hard, adherent deposits. Most magnesium phosphate deposits are colorless but become red, brown, or black when contaminated with iron oxides.

Insoluble silicates are present in many boilers. Many silicates are very hard and are almost insoluble in acids except for hydrofluoric. Complex silicates such as analcite ($Na_2O \cdot Al_2O_3 \cdot 4SiO_2 \cdot 2H_2O$), acmite

Figure 1.17 Calcium carbonate deposit (see Fig. 1.10).

Figure 1.18 Analcite formed by carryover of alum from clarifiers. Note how the deposit has spalled from the bulge.

$(Na_2O \cdot Fe_2O_3 \cdot 4SiO_2)$, or sodalite $[Na_2O \cdot 3Al_2O_3 \cdot SiO_2 \cdot 2NaCl]$ can be evidence for steam blanketing, since intermediate sodium compounds are soluble. Analcite frequently forms if alum is carried over from clarifiers (Fig. 1.18).

In general, water-soluble deposits can be retained only if localized concentration mechanisms are severe. Therefore, the presence of species such as sodium hydroxide, sodium phosphate, and sodium sulfate should be considered proof of evaporation to dryness. It has been reported that the strong odor of hydrogen sulfide due to sulfite decomposition was detected near a boiler where hot spots were present.

Hydrolyzable salts such as $MgCl$ can gain entrance to boiler water through leaks in seawater condensers. Chlorides will concentrate in porous deposits and hydrolyze to form hydrochloric acid. Decomposition of oil, grease, and other organic contaminants can also produce acid conditions.

Deposit morphology

Much of the preceding discussion involves the identification of deposits based only on individual chemical natures. However, most real boiler deposits contain many intermixed compounds. The deposit morphology can influence the species which are found and also gives clues as to the heat input, water conditions, and corrosion mechanisms under which they were formed.

Any deposit that resists rinsing is especially undesirable. In particular, insoluble, permeable deposits such as coarse particulate matter are harmful. They permit the concentration of more soluble species in their pores. Hence, sodium hydroxide may be found intermixed with porous, insoluble species. Deposition of the porous species often predates the concentration of the more soluble species.

In general, the older the deposit, the harder it is and the more tenaciously it adheres to tube walls. As deposits age, they fill their interstices with solid material, resulting in increased bulk, density, and hardness. There are exceptions to this rule, and these judgments are naturally somewhat subjective. Nevertheless, the presence of hard, tenacious deposits often indicates that deposition has been occurring for a long time at high metal temperatures.

Deposit stratification indicates changes in water chemistry, heat input, etc. It is often possible to identify deposits formed during different water-treatment programs through careful inspection and analysis of laminate scales. The order in which the deposits appear usually can provide a chronological deposition history.

Elimination

All deposits are undesirable and ultimately result from water-chemistry properties or boiler operations practices. Proper water treatment can reduce deposition. The general rules of proper water treatment are obvious. Water-chemistry upsets and operation changes should be minimized.

Deposition can be avoided by operating at or slightly below design loads and ensuring that all boiler components are functioning properly. The most important boiler operating characteristic influencing deposition is firing practice. Also, elimination of hot spots, correct monitoring of water levels, and maintenance of a constant load are necessary to avoid deposition. In addition, correct burner position, well-considered fuel adjustments, and appropriate blowdown practices contribute to reduced deposition.

Cautions

Deposits rarely contain only a single compound. Chemical analysis is often necessary to determine the amount and variety of each chemical species. Soluble species will be washed away when the boiler cools and steam generation ceases. Washing sometimes results in laboratory analysis that does not accurately reflect in-service compositions. It is usually safe to assume that concentrations of highly soluble species are underreported in laboratory analysis. The presence of any highly soluble material is usually sufficient to prove deviations from nucleate boiling.

The amount, composition, and stratification of deposits are often altered near a rupture site. Escaping fluids may remove deposits near the rupture. Occasionally, fire-side combustion products may find their way onto internal surfaces near the rupture. Weight determinations for deposits near bulged or ruptured surfaces are usually underestimates of actual in-service values.

Corrosion products can be confused with deposits. This is especially true in superheaters suffering idle-time oxygen corrosion, where tubercular growth is often confused with carryover of boiler-water solids (see Chap. 8, "Oxygen Corrosion"). Magnetite needles associated with low-pH excursions can also be confused with particulate iron brought in from the preboiler system.

Related Problems

See also Chap. 2, "Long-Term Overheating"; Chap. 3, "Short-Term Overheating"; Chap. 4, "Caustic Corrosion"; Chap. 6, "Low-pH Corrosion during Service"; Chap. 7, "Low-pH Corrosion during Acid Cleaning"; and Chap. 8, "Oxygen Corrosion."

CASE HISTORY 1.1

Industry:	Pulp and paper
Specimen Location:	Economizer tube from recovery boiler near upper header
Orientation of Specimen:	Curved (horizontal to vertical — J-shaped)
Years in Service:	10
Water-Treatment Program:	Chelant
Drum Pressure:	600 psi (4.13 MPa); feedwater pressure, 820 psi (5.65 MPa)
Tube Specifications:	2 in. outer diameter
Fuel:	Black liquor

During removal of failed economizer tubes, heavy internal deposits were found unexpectedly. Internal surfaces of some economizer tubes were partially lined with an irregular layer of soft, flaky iron oxide. Underlying oxides were black. A 1-in.-thick layer of friable material accumulated at the tube bend is shown in Fig. 1.19. Close observation revealed that the accumulated material consisted of anthracite particles and resin beads (Fig. 1.20). The deposit had been transported from a disintegrated resin bed. The metal surface beneath the deposit was gouged and thinned.

The accumulated deposit retarded coolant flow, yet overheating could not occur because flue-gas temperatures were too low (about 550°F, or 288°C). Deviations from nucleate boiling beneath the deposit did occur, leading to concentration of sodium hydroxide and associated corrosion (see Chap. 4, "Caustic Corrosion").

It is extraordinary that this deposit did not cause failure. Rather, other

Figure 1.19 Anthracite and resin beads beneath a stratified iron oxide layer.

Figure 1.20 Anthracite and resin beads. (Magnification: 7.5X.)

nearby tubes failed as a result of erosion from external surfaces and oxygen corrosion. At this time the deposits were discovered.

The resin-bed disintegration was not found until the economizer deposits were found. Although these deposits must have been carried into other parts of the boiler, no failures were traced to their presence.

CASE HISTORY 1.2

Industry:	Pulp and paper
Specimen Location:	Superheater outlet
Orientation of Specimen:	Vertical (stub end down)
Years in Service:	3½
Water-Treatment Program:	Phosphate
Drum Pressure:	900 psi
Tube Specifications:	2 in. (5.1 cm) outer diameter, stub-end, low-alloy steel (1½% Cr)
Fuel:	Coal

A 2-ft section of superheater-tubing stub end was plugged solid with deposits that were still soaked with water when the tube was split. Sections were gently heated for 3 days to dry the deposit in situ. After drying, white deposits became visible at the end of the stub (the lowest tube point in service) (Fig. 1.21).

The tube contained about 2 lb of deposit for every 2 in. (5.1 cm) of its length. Away from the stub end, the deposit contained about 80% magnetite

Figure 1.21 Longitudinally split stub end from a superheater outlet. The tube is plugged with deposits. The white material is sodium hydroxide and sodium carbonate concentrated at the bottom of the stub end. The black material is mostly magnetite.

and 7% sodium hydroxide and sodium carbonate by weight. Small amounts of sulfur, chlorine, phosphorus, chromium, and manganese were also detected. At the stub end, about half the deposit was sodium hydroxide and sodium carbonate.

Deposits were caused by chronic contamination of attemperation water and/or carryover of boiler water into the superheater header from the steam drum. Water was converted to steam and left solids behind. Solids accumulated during a considerable period in stagnant, low-flow regions like stub ends.

The color difference between deposits at the stub end and shank indicates segregation of the soluble white alkaline material to the lowest portion of the stub. Segregation is caused by percolation of solute-rich water to the stub end. The fact that the white material was concentrated at the bottom of the tube after drying supports this conclusion.

Cleaning these stub ends is a very difficult task. However, cleaning or removal of the ends is necessary to prevent caustic corrosion and overheating.

CASE HISTORY 1.3

Industry:	Pulp and paper
Specimen Location:	Crossover tube
Orientation of Specimen:	Horizontal
Years in Service:	20
Water-Treatment Program:	Phosphate
Drum Pressure:	800 psi (5.52 MPa)
Tube Specifications:	2½ in. (6.4 cm) outer diameter
Fuel:	Black liquor and fuel oil

The crossover tube contained a small bulge and rupture. A white, friable deposit layer having a density of 16 g/ft^2 (17 mg/cm^2) was present on hot-side internal surfaces; 1 g/ft^2 was present on the cold side. Deposits were bounded on both sides by sharp borders (Fig. 1.22). The borders indicate the steam-blanketing terminus. Deposition took place during a brief episode of high fire-side heat input. High heat input resulted in overheating and eventual failure.

Figure 1.22 Deposit layer on the internal surface of the hot side of a crossover tube. Some deposit spalled off at a shallow bulge. Note the sharp boundaries indicating the terminus of steam blanketing.

Deposits consist of basic calcium phosphate (hydroxyapatite), sodium aluminum silicate hydrate, magnetite, magnesium silicate, and small amounts of other materials. Some alum was carried over from clarifiers. Firing practices and water clarification were reviewed and changed appropriately.

CASE HISTORY 1.4

Industry:	Pulp and paper
Specimen Location:	Primary superheater, pendant U-bend
Orientation of Specimen:	Vertical
Years in Service:	11
Water-Treatment Program:	Phosphate
Drum Pressure:	1250 psi (8.62 MPa)
Tube Specifications:	1¾ in. (4.45 cm) outer diameter; 2¼% Cr, 1% Mo
Fuel:	Hog fuel and coal

A pendant U-bend section of the superheater was removed from service because of fire-side metal loss associated with slagging. When the internal surface was examined, severe blistering of the magnetite layer was found (Fig. 1.23). Spalling of the magnetite layer had occurred. Although such spalling can lead to turbine erosion, none was detected.

Oxide spalling (exfoliation) is caused by fluctuating stresses, usually

Figure 1.23 Magnetite blisters on the internal surface of a U-bend in a primary superheater. Note that the magnetite has spalled in places.

fluctuating thermal stresses. Stresses can become large when tubes are slagged. Furnace-gas channeling and "laning" can produce hot spots. The fracturing of slag, perhaps during soot blowing, can cause rapid surface-temperature variations. It is not established that soot blowing alone can cause exfoliation.

Magnetite bubbles grow because the thermally formed magnetite occupies a larger volume than the metal from which it forms. Therefore, compressive stresses may occur in the growing layer, tending to cause buckling. When temperatures are high, the oxide may slowly deform instead of fracturing.

Virtually all superheaters experience some magnetite exfoliation as they age. However, slagging, large load swings, mechanical vibration, and frequent up-and-down operation accelerate the stresses leading to exfoliation.

CASE HISTORY 1.5

Industry:	Utility
Specimen Location:	Fine turbine screen
Orientation of Specimen:	Vertical
Years in Service:	8
Water-Treatment Program:	Coordinated phosphate
Drum Pressure:	2200 psi (15.2 MPa)
Sample Specifications:	0.035-in. (0.088-cm) mild steel wire; screen-opening size of 0.040 in. (0.012 cm)
Fuel:	Coal

After routine superheater inspections, it was revealed that spalling and exfoliation of magnetite on internal surfaces had occurred. Sections of both coarse and fine turbine-inlet screens were removed for inspection.

Approximately 10% of the fine screen openings were plugged with magnetite chip scale. Small exfoliated chips had been blown through the superheaters and had lodged in the screen (Fig. 1.7). A few rocks and cutting-torch spatter beads were also trapped. The coarse screen caught magnetite chips as thick as 0.010 in. (0.025 cm) and as wide as 0.15 in. (0.038 cm). The side of the coarse screen facing steam flow was hammered and peened by impact with flying projectiles.

A slight decrease in turbine efficiency had been noticed recently, and it was suspected that hard-particle erosion from exfoliated oxide might be damaging buckets. Subsequent turbine inspections revealed only minor damage.

Chemical treatments to reduce slagging were suggested, and soot-blowing practices were reviewed. Chemical cleaning of the superheater was deemed unnecessary at the time.

CASE HISTORY 1.6

Industry:	Chemical plant
Specimen Location:	Target-wall tube, 6 ft (1.8 m) above lower-wall header
Orientation of Specimen:	Horizontal
Years in Service:	5
Water-Treatment Program:	Polymer
Drum Pressure:	425 psi (2.93 MPa)
Tube Specifications:	3¼ in. (8.26 cm) outer diameter
Fuel:	Natural gas and No. 6 fuel oil

One of the seventy-four target-wall tubes was bulged and ruptured. A thick, friable, tan deposit layer covered the internal surface (Fig. 1.24). The hot-side deposit weight was 100 g/ft², while the cold-side weight was 82 g/ft². Deposits contained calcium carbonate, calcium silicate, calcium phosphate, and magnesium silicate hydroxide.

The boiler was fired intermittently at up to 90,000 pounds (40,900 kilograms) of steam per hour for over 1 year [design was 75,000 lb/h (34,000

Figure 1.24 Cross section of bulged target-wall tube. Deposits are heavy on both hot and cold sides. Most deposits are phosphates, silicates, and carbonates.

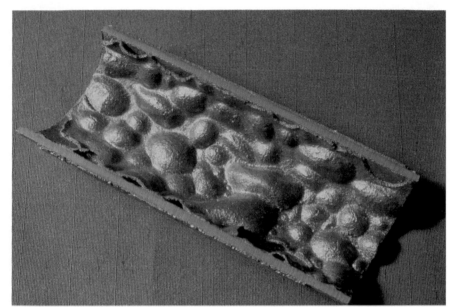

Figure 2.4 Thick, blistered magnetite layer on internal surface of 150-psi (1.0-MPa) tube. Oxide was formed during dry firing of boiler. Metal temperatures approached 2000°F (1100°C), causing slow, "viscous" flow of magnetite.

general, metal temperature is higher at bulges than in the surrounding metal. In many cases, continued existence of bulges is sufficient to cause overheating, even if associated deposits are removed.

Thermal oxidation (metal burning)

One sign of long-term overheating can be a thick, brittle, dark oxide layer on both internal and external surfaces (Figs. 2.3 and 2.4). If metal temperatures exceed a certain value for each alloy, thermal oxidation will become excessive. The temperatures at which alloys appreciably oxidize (scaling temperatures) are given in Table 2.1. Often, the thermally formed oxide layer contains longitudinal fissures and cracks. In other areas, patches of oxide may have exfoliated (Fig. 2.5). Cracks and exfoliated patches result from tube expansion and contraction caused by deformation during overheating and/or thermal stressing. Tube-wall thinning can result from cyclic thermal oxidation and spalling. This process can continue until the entire wall is converted to oxide, creating a hole (Fig. 2.6).

Creep rupture (stress rupture)

Creep rupture (stress rupture) is a form of long-term overheating damage that usually produces a thick-lipped rupture at the apex of a bulge. Creep produces slow plastic deformation and eventual coalescence of microvoids

Figure 2.5 Thermally deteriorated metal on failed wall tube of a utility boiler. Note the spalled and cracked oxide resembling tree bark caused by expansion of tube during bulging and thermally induced stresses.

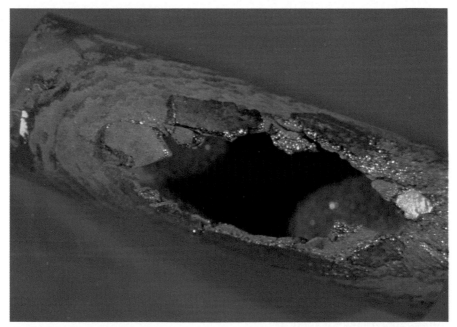

Figure 2.6 Through-wall oxidation of a tube from a low-pressure boiler. Brittle black magnetite patches still adhere to the rupture mouth. Elsewhere the oxide has spalled. Such through-wall oxidation is most common in low-pressure boilers, where internal pressures are not sufficient to cause premature failure.

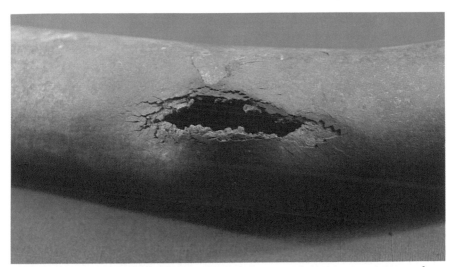

Figure 2.7   A small, ragged creep rupture at the apex of a bulge. Note the small secondary fissures and the relatively thick rupture edges.

in metal during overheating. Often a small longitudinal fissure will be present at the apex of a heavily oxidized bulge (Fig. 2.7). In other cases, the rupture will be larger and have a fish-mouth shape (Fig. 2.8). The rupture will usually have blunt and slightly ragged edges. Similar, but smaller, longitudinally oriented ruptures and fissures may exist nearby.

Chain graphitization

A somewhat uncommon form of damage, *chain graphitization,* can also occur after long-term overheating. The damage begins when iron carbide

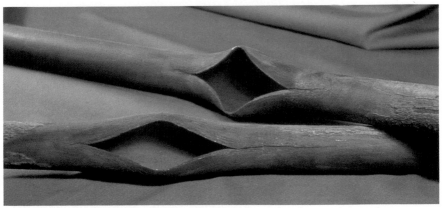

Figure 2.8   Two superheater tubes from a utility boiler [1400 psi (9.7 MPa)], failed by creep. Note the fish-mouth ruptures. Compare with Fig. 2.7.

particles (normally present in plain carbon or low-alloy steels) decompose into graphite nodules after prolonged overheating above 800°F (427°C). The graphite nodules, if distributed uniformly in the steel, rarely cause failure. However, nodules sometimes chain together, forming planes of cavities filled with graphite. The nodules usually form at microstructure defects, in places where there are chemical impurities, and along stress lines. Stresses generated by internal pressures cause tearing of the metal along chains of nodules, much as a postage stamp is torn from a sheet along the perforated edges.

Chain graphitization is usually found at welds (Fig. 2.9). Less commonly, damage occurs away from welds and forms helical cracks spiraling around tube surfaces (Fig. 2.10). Such failures are sometimes confused with creep damage, but careful microscopic observation will reveal the presence of graphite nodules at or near fracture edges.

Critical Factors

Long-term overheating is a chronic, rather than transient, problem. It is the result of long-term deposition and/or long-term system operating problems. Heavy internal deposition on both hot and cold sides of water-

Figure 2.9 Rupture at longitudinal weld due to chain graphitization. Note the blunt, rough rupture edge. Such failures occur after long-term mild overheating.

Figure 2.10 Spiral through-fissures due to chain graphitization. These fissures formed along lines of maximum shear stress.

wall tubes often indicates that deposits have insulated the tube wall from cooling effects of the water, contributing to overheating.

Experience has shown that when the ratio of deposits on hot to cold faces of water-cooled tubes exceeds 3, fire-side heat input is substantially higher on the hot face. When this ratio approaches 10, heat input on the hot face relative to the cold face can be quite excessive. In many cases, the ratio will be below 3 if the heat input is not excessive on the hot face; when chemical water treatment is deficient; or when water chemistry is overwhelmed by contaminants.

Deposits on superheater tubes caused by carryover and/or contaminated attemperation waters can produce overheating. In most steam-cooled tubes, heat-flux differences between the hot and cold sides are not pronounced. Resulting deposit patterns are characteristic of contamination rather than excessive heat input. Other sources of overheating include overfiring, incorrect flame pattern, restricted coolant flow, inadequate attemperation, and improper alloy composition.

Identification

A thick, brittle magnetite layer near the failure indicates long-term overheating. At greatly elevated temperatures (short-term overheating) reduction in metal strength is such that failure occurs before significant amounts of oxides can develop.

Bulging and plastic deformation are almost always present if the tube is pressurized. Tubes ruptured as a result of long-term overheating usually show bulging and plastic deformation at the failure site. The rupture is almost always longitudinal, with a fish-mouth shape. Rupture edges may be knifelike or thick depending on the time, temperature, and stress levels involved. Multiple bulges may occur.

Water-side deposits will usually be present and will often be hard and stratified. Deposits will usually be "baked" onto the wall and will become hard and brittle. Deposits tend to show multiple layers of different colors and textures, the innermost layers being hardest and most tenacious.

Visual inspection is adequate for oxidation, spalling, and bulging. Thermocouple measurement during service frequently supplies good information. The best way to ascertain that long-term overheating has occurred is by metallographic inspection of a failed tube.

Elimination

Eliminating long-term overheating requires removal of a chronic system defect. Headers, U-bends, long horizontal runs, and the hottest areas should be inspected for evidence of obstruction, scales, deposits, and other foreign material. Excess deposits should be removed by chemical or mechanical cleaning and prevented from recurring. Firing procedures, Btu value of fuels, and in-service furnace temperatures near overheated areas should be checked. Attemperation procedures should also be reviewed. If needed, changes in metallurgy, tube shielding, and the judicious use of refractories should be considered.

The source of significant deposits must be identified and eliminated. Common causes of deposits include improper water treatment, system contamination, improper boiler operation, and/or excessive heat input. Each potential cause must be addressed methodically.

Cautions

It is incorrect to assume that long-term overheating automatically produces significant tube damage. While small microstructural changes may occur in the tube wall, these changes often do little to reduce service life or significantly weaken the tube. However, if overheating continues for a long time, failures will eventually result.

Long-term and short-term overheating failures may appear similar. Frequently, evidence of both long-term and short-term overheating will be present at the same failure. Failures due to long-term overheating are sometimes associated with chemical attack and other significant metal wastage, while chemical attack in short-term overheating is rare. The presence of corrosion does not exclude consideration of long-term overheating. Because a brief episode of short-term overheating may follow long-term overheating, the sequence of thermal events can be difficult to diagnose without microscopic examinations.

Related Problems

See also Chap. 1, "Water-Formed and Steam-Formed Deposits"; Chap. 3, "Short-Term Overheating"; Chap. 4, "Caustic Corrosion"; and Chap. 14, "Hydrogen Damage."

CASE HISTORY 2.1

Industry:	Gas products
Specimen Location:	Superheater, 3 ft (1 m) above firebox floor near the center of the boiler
Specimen Orientation:	Horizontal, immediately adjacent to a firebrick wall
Years in Service:	12
Water-Treatment Program:	Phosphate
Drum Pressure:	Design is 700-psi (4.8-MPa) package boiler, but operated at 600 psi (4.1 MPa)
Tube Specifications:	2 in. (5.1 cm) outer diameter, SA-213-T22
Fuel:	Natural gas

A brittle black magnetite layer covered both external and internal surfaces of this superheater tube. The thermally formed oxide fractured and spalled, substantially reducing wall thickness (Fig. 2.11). Thinning was more severe along the side of the section abutting a firebrick wall.

Most metal was lost from external surfaces and was caused by thermal oxidation at temperatures between 1100 and 1350°F (600 and 727°C). Gas channeling and higher temperatures were present along the tube side abutting the firebrick. These factors accelerated oxidation and spalling processes.

In the past, nearby tubes had ruptured as a result of thinning associated with thermal oxidation. No deposits were present on internal surfaces, and attemperation was not used. Fire-side slagging was not present in this intermittently run boiler.

Failures were a chronic problem associated with design and operation. The proximity of failures to the firebrick wall strongly linked the overheating to boiler design.

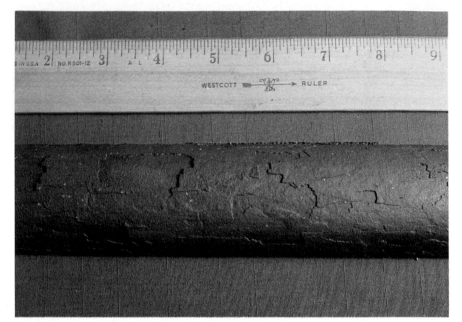

Figure 2.11 Severe thermal oxidation and spalling of magnetite on external surface. The laminated nature of the scale indicates multiple episodes of spalling and oxide reformation.

CASE HISTORY 2.2

Industry:	Steel
Specimen Location:	Center of waterwall
Specimen Orientation:	Vertical
Years in Service:	8
Water-Treatment Program:	Phosphate
Drum Pressure:	1200 psi (8.3 MPa)
Tube Specifications:	3 in. (7.6 cm) outer diameter
Fuel:	Blast-furnace gas

Longitudinal fissures and thick-walled, ruptured bulges were present along the hot side of this section (Fig. 2.12). Surfaces are covered with an irregular tan slag layer, and the external surface near the through-fissure is checkered.

The rupture and fissuring were caused by very long-term exposure of the metal to temperatures between 850 and 1050°F (454 and 566°C). Evidence suggests that exposure to these temperatures may have been occurring for several years. Overheating was caused by excessive heat input relative to coolant flow rate.

Figure 2.12 Longitudinal fissures along hot side. Note the through-fissure at the bulge apex.

Nearby tubes showed similar, although less severe, attack. Deposits on hot-side internal surfaces were less than 20 g/ft^2 (22 mg/cm^2). The boiler had been cleaned 3 years prior to failure.

CASE HISTORY 2.3

Industry:	Utility
Specimen Location:	Rear wall, 30 ft (9.1 m) from front wall
Specimen Orientation:	Vertical
Years in Service:	1½
Water-Treatment Program:	Polymer
Drum Pressure:	620 psi (4.3 MPa)
Tube Specifications:	3 in. (7.6 cm) outer diameter
Fuel:	Pulverized coal

The tube has a series of several prominent bulges longitudinally aligned along the hot-side crown (Fig. 2.2). Thick layers of hard, tenacious iron oxide cover each bulge, except where spalling has dislodged the oxide. Internal surfaces on the hot side are covered with spongy, porous deposits, which cover a hard, black magnetite layer. The back wall at the same elevation saw many failures. The boiler had frequent load swings and was operated intermittently.

The tube was overheated in a temperature range between 950 and 1150°F (510 and 620°C) at the bulges for a long time. Formation of deposits was due to an imbalance between coolant flow and fire-side heat input. Deposit weights were about 5 g/ft² (5 mg/cm²) on the cold side and 26 g/ft² (28 mg/cm²) on the hot side. Deposits were caused by exceeding the solubility of inversely soluble species, by evaporative concentration, and by mechanical entrainment of particulate matter.

Because of the frequent load swings and intermittent operation, closer operational monitoring was suggested.

CASE HISTORY 2.4

Industry:	Steel
Specimen Location:	Primary-superheater bank
Specimen Orientation:	Vertical pendant
Years in Service:	16
Water-Treatment Program:	Polymer
Drum Pressure:	1200 psi (8.3 MPa)
Tube Specifications:	2¼ in. (5.7 cm) outer diameter
Fuel:	Coke gas

Three large bulges are present on the pendant legs of this U-bend section. Each bulge is ruptured at its apex. The legs expanded to about a 2½-in.

Figure 2.13 Bulged and ruptured superheater U-bend. The legs are slightly expanded.

Figure 2.14 As in Fig. 2.13. Fragmented thermally formed oxide surrounds the rupture.

(6.4-cm) outer diameter from their 2¼-in. (5.7-cm) original value. Bulges are shown in Figs. 2.13 and 2.14. Internal surfaces are free of significant deposits, while external surfaces are covered with a tenacious, fragmented oxide layer. Eighteen tubes were similarly affected. This boiler operates on waste heat from a slab furnace.

The tube was overheated for a period of several days or longer at temperatures above 1000°F (540°C), but below 1350°F (730°C). The wall strength was decreased at elevated temperatures, and tubes were thinned and weakened by thermal oxidation. As a result of these two forms of weakening, the legs bulged and then ruptured.

Operating records showed an increase of almost 100°F (56°C) in steam temperature approximately 2 months prior to failure. This increase was correlated with changes in firing associated with alteration of the slab mill's operation.

CASE HISTORY 2.5

Industry:	Utility
Specimen Location:	Primary-superheater inlet; section with hottest flue gas
Specimen Orientation:	Horizontal
Years in Service:	20
Water-Treatment Program:	Phosphate
Drum Pressure:	1200 psi (8.3 MPa)
Tube Specifications:	Circumferentially welded tubes, 2¼ in. (5.7 cm) outer diameter
Fuel:	Currently coal (1% S, less than 10% ash); until 6 months before failure, oil (2% S, 200–400 ppm V) was used as fuel

The section contains a massive, thick-walled, longitudinal rupture just above a circumferential weld. The rupture occurred immediately downstream of the weld. The tube was bent into an "L" by the burst. The rupture terminates at one end in a pair of thick-walled transverse tears (Fig. 2.15). A thick, tenacious magnetite layer covers external surfaces, except those near the

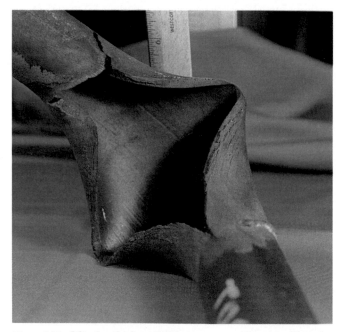

Figure 2.15   Massive thick-walled fracture caused by creep. Note the circumferential weld just below the failure. The tube is torn and bent to a 90° elbow by the violence of the rupture.

rupture, where the oxide has cracked and spalled. Elemental copper and spotty deposits are present on internal surfaces.

Failure was caused by prolonged overheating at temperatures above 1050°F (570°C). The direct rupture was a stress (creep) rupture. Coolant flow irregularities immediately downstream of a partially intrusive circumferential weld, along with internal deposition, which reduced heat transfer, were contributing factors. Additionally, a switch from oil to coal firing likely changed fire-side heat input.

The superheater had a history of boiler-water carryover, and load swings were common. Previous failure had occurred in this region at least 2 years before the current rupture.

Short-Term Overheating

Locations

Failures caused by short-term overheating are confined to steam- and water-cooled tubes including downcomers, waterwalls, roofs, screens, superheaters, and reheaters. Because of their high operating temperatures, superheaters and reheaters are common failure sites. Failures due to short-term overheating almost never occur in economizers, where temperatures are limited.

When low water level is the cause, failures will often occur near the top of waterwalls near steam drums. A single ruptured tube in the midst of other apparently unaffected tubes suggests pluggage or other flow-related problems.

Poor attemperation usually will not cause short-term overheating, although long-term overheating may occur. Failures of superheaters and reheaters can also occur during start-up, when steam flow is limited.

General Description

Short-term overheating occurs when the tube temperature rises above design limits for a brief period. In all instances, metal temperatures are at least 850°F (454°C) and often exceed 1350°F (730°C). Depending on temperature, failure may occur in a very short time. Failure is usually caused by a boiler operation upset. Conditions leading to short-term overheating are partial or total tube pluggage and insufficient coolant flow due to upset conditions and/or excessive fire-side heat input.

Critical Factors

An occurrence of short-term overheating is caused by an unusual set of circumstances, such as an upset, occurring during a brief period. Therefore, pinpointing unusual events immediately preceding failure may be extremely important in identifying the cause of failure.

Since short-term overheating frequently has little to do with water chemistry, efforts should be concentrated on operating procedures and system design. Did failure occur on start-up or shutdown? Was there a recent acid cleaning? Were headers or U-bends filled with debris upstream of the failure? Did another failure immediately precede this one? Was the firing pattern changed? Was there an unusually large load swing? Was water level normal? Was blowdown unusually severe before or during failure?

Identification

Short-term overheating frequently can be identified by metallographic examination. Such analysis requires sectioning of the tube for microscopic examination. Most other identification techniques are less effective.

Several factors often present in failures caused by short-term overheating are uniform tube expansion, absence of significant internal deposits, absence of large amounts of thermally formed magnetite, and violent rupture.

Short-term overheating may produce bulging. In very rapid overheating, a thick-walled longitudinal rupture (Fig. 3.1), or a longitudinal fish-mouth rupture (Fig. 3.2), can occur.

At elevated temperatures metal strength is markedly reduced (Fig. 2.1). In fact, if temperatures rise to very high levels, failure will occur quickly. If failure happens rapidly, bulging may be absent and the rupture can be violent, sometimes bending the tube almost double and causing secondary metal tearing (Fig. 3.3). In the case of thick-walled ruptures, the tube circumference at the rupture is sometimes nearly exactly equal to the tube circumference away from the rupture. Tube circumference can be roughly measured with a piece of string.

Rupture edges may be blunt and retain most of their original wall thickness or gradually taper to knifelike or chisellike edges. In some cases, the tube diameter may be uniformly expanded with no rupture occurring.

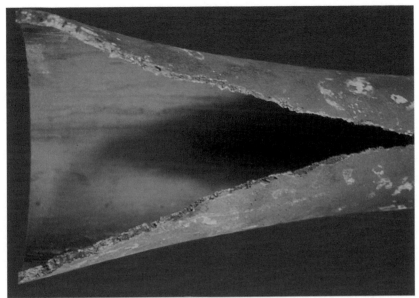

Figure 3.1 Longitudinal rupture in a superheater tube caused by partial pluggage upstream of failure, which in turn caused short-term overheating. Note the thick-walled rupture edges. Virtually identical tube circumferences are present at the rupture and away from the burst. Such failures often occur when temperatures exceed 1350°F (730°C).

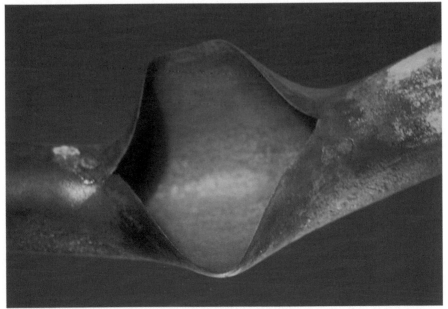

Figure 3.2 Short-term overheating in which bulging occurred before rupture. Note the chisel-like rupture edges.

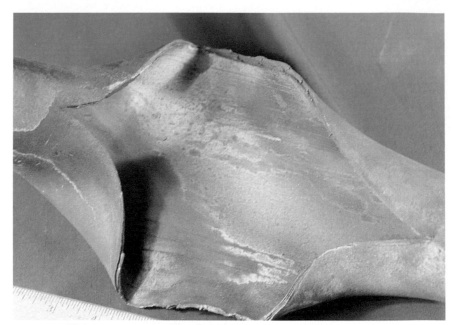

Figure 3.3 Violent rupture caused by short-term overheating. The tube is bent almost to a right angle, caused by the severity of the burst.

Multiple bulges are usually absent, although a single bulge containing a rupture may occur, especially if long-term overheating has occurred previously (Fig. 3.4).

In general, heavy internal deposits will not be present in a short-term rupture since these deposits are not likely to be the cause of the rupture. Further, if deposits do occur, they usually will be friable and easily removed by gentle probing, rather than baked on to the surface as is typical in long-term overheating. Thick accumulations of thermally deteriorated metal will be absent.

Elimination

The solution of short-term overheating, which is often caused by brief upset conditions, is to eliminate the upset. If restricted coolant flow due to tube pluggage is suspected, drums, headers, U-bends, long horizontal runs, and other areas where debris may accumulate should be inspected and cleaned. This is especially true if a failure occurs shortly after a boiler cleaning. The drum water level, firing procedures, and blowdown and start-up procedures should be carefully monitored. Suspicion should be aroused if a short-term failure occurs immediately after another failure. A

Figure 3.4 A rupture at a single bulge. The tube had experienced long-term overheating, followed by a brief episode of severe overheating.

failure may disturb circulation, dislodge deposits, and dislodge corrosion products. This, in turn, may affect heat transfer.

Cautions

A thick-walled longitudinal rupture by itself is not sufficient to warrant a diagnosis of short-term overheating. Be extremely wary if the rupture occurs at a weld or a tube seam or if any evidence of metal wastage occurs near the failure. Thick-walled failures due to short-term overheating can easily be confused with failures from long-term overheating involving creep (stress rupture), failures from hydrogen embrittlement, and failures from certain tube defects. The absence of deposits near a rupture may be due to the scrubbing action of escaping fluids during rupture. Also, short-term overheating may occur after long-term overheating. Such involved mechanisms usually require metallurgical examination to determine failure modes.

Microstructural changes occurring during short-term overheating do not always lead to failure. In addition, tubes that have experienced short-term overheating do not always have to be replaced. Mechanical properties are not necessarily altered significantly by the overheating.

Related Problems

See Chap. 2, "Long-Term Overheating."

CASE HISTORY 3.1

Industry:	Utility
Specimen Location:	Waterwall, nose arch
Specimen Orientation:	Slanted
Years in Service:	5
Water-Treatment Program:	Coordinated phosphate
Drum Pressure:	1800 psi (12.4 MPa)
Tube Specifications:	2½ in. (6.4 cm) outer diameter, internally rifled
Fuel:	Pulverized coal

This section of the internally rifled waterwall nose arch contains a large, thick-wall rupture. Rupture edges are 12 in. (30 cm) long and have a jagged contour (Fig. 3.5). Both internal and external surfaces are smooth and are

Figure 3.5   Large, fish-mouth rupture of rifled nose-arch tube. Rupture edges are thick, blunt, and ragged. Note the absence of significant deposits.

covered with thin, tenacious, dark oxide layers. No significant deposits are present anywhere.

Rupture occurred shortly after start-up. Microstructural evidence indicated that the tube metal near the rupture exceeded 1600°F (870°C). No significant accumulation of thermally formed oxide was found anywhere on the received section.

Internal rifling is sometimes used to reduce steam channeling and inhibit steam blanketing. No steam blanketing or liquid/vapor interface was found on internal surfaces. There was no change in microstructure on the cold side indicating that the tube contained water at the time of the rupture. Rather, the burst was caused by insufficient coolant flow on start-up.

CASE HISTORY 3.2

Industry:	Refinery
Specimen Location:	Powerhouse, target tube near steam drum
Specimen Orientation:	Slightly curving from vertical
Years in Service:	25
Water-Treatment Program:	Phosphate
Drum Pressure:	800 psi (5.5 MPa)
Tube Specifications:	3¼ in. (8.3 cm) outer diameter

The boiler from which the tube was removed has been subject to wide load swings since the installation of new major waste-heat generators. The boiler often remains on "ready standby," where it is kept either on very low fire or off-line. Low-pressure steam is returned to the steam drum for some heat (and no circulation).

A massive thin-lipped longitudinal rupture is present. A thin, straight through-fissure runs longitudinally down the tube from the corner of the rupture mouth (Fig. 3.6). Internal surfaces are smooth, except for shallow mandrel markings. A thin, uniform layer of light-colored deposits is present away from the rupture.

Failure was caused by severe overheating resulting from coolant starvation. A minor defect (mandrel marking) guided the rupture line but was not a significant contributing factor to failure.

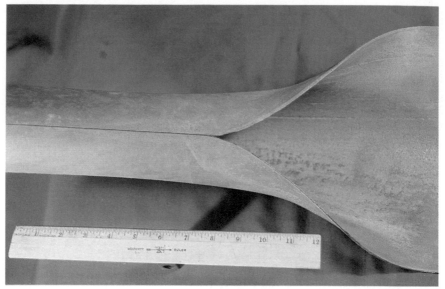

Figure 3.6 Longitudinal fissure running from the edge of a large, wide-open rupture.

CASE HISTORY **3.3**

Industry:	Utility
Specimen Location:	Superheater from a waste-heat boiler
Specimen Orientation:	Vertical
Years in Service:	10
Water-Treatment Program:	Polymer
Drum Pressure:	600 psi (4.1 MPa)
Tube Specifications:	3 in. (6.6 cm) outer diameter with spirally wound $^7/_{16}$-in. (1.1-cm) by $^3/_{16}$-in. (0.5-cm) fins
Fuel:	Waste heat from gas turbines

A rupture 4¼ in. (10.8 cm) long by 1⅝ in. (4.1 cm) wide is present at a bulge (Fig. 3.7). A thick, black layer of thermally formed oxide is present on both internal and external surfaces and is thicker along the hot side. Deposits containing sodium, calcium, silicon, and iron are present in the tube. Deposit loading is 38 g/ft² (41 mg/cm²). Carryover of boiler water into the superheater due to water-level excursions and foaming was common. The boiler had never been cleaned.

Temperatures of waste-heat gases ranged from 1200 to 1800°F (650 to 980°C). The tube failed during several other tube failures. Tubes nearby

Figure 3.7 Large, wide-open rupture in a superheater tube of a waste-heat boiler. The tube experienced long-term overheating followed by a brief episode of short-term overheating when nearby tubes ruptured.

failed as a result of long-term overheating (creep rupture), while others failed as a result of short-term overheating. It was determined that the section had been mildly overheated for a considerable period, but had failed during a brief episode of severe overheating when temperatures exceeded 1350°F (730°C). The short-term overheating probably occurred because of coolant starvation, caused by leakage from other failed tubes upstream.

Caustic Corrosion

Locations

Generally, caustic corrosion is confined to (1) water-cooled tubes in regions of high heat flux, (2) slanted or horizontal tubes, (3) locations beneath heavy deposits, and (4) heat-transfer regions at or adjacent either to backing rings at welds or to other devices that disrupt flow.

General Description

The terms *caustic gouging* and *ductile gouging* refer to the corrosive interaction of sufficiently concentrated sodium hydroxide with a metal to produce distinct hemispherical or elliptical depressions. The depressions may be filled with dense corrosion products that sometimes contain sparkling crystals of magnetite. At times, a crust of hard deposits and corrosion products containing magnetite crystals will surround and/or overlie the attacked region. The affected metal surface generally has a smooth, rolling contour.

The susceptibility of steel to attack by sodium hydroxide is based on the amphoteric nature of iron oxides; that is, oxides of iron are corroded by both low-pH and high-pH environments (Fig. 4.1). High-pH substances, such as sodium hydroxide, will dissolve magnetite:

$$4NaOH + Fe_3O_4 \rightarrow 2NaFeO_2 + Na_2FeO_2 + 2H_2O$$

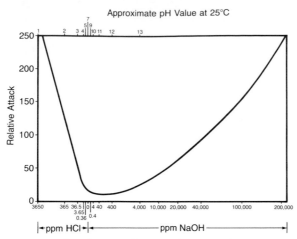

Figure 4.1 Attack on steel at 310°C (590°F) by water of varying degrees of acidity and alkalinity. (*Curve by Partridge and Hall, based on data of Berl and van Taack. Courtesy of Herbert H. Uhlig,* The Corrosion Handbook, *John Wiley and Sons, New York, 1948.*)

When magnetite is removed, the sodium hydroxide may react directly with the iron:

$$Fe + 2NaOH \rightarrow Na_2FeO_2 + H_2$$

Critical Factors

Two critical factors contribute to caustic corrosion. The first is the availability of sodium hydroxide or of alkaline-producing salts (i.e., salts whose solution in water may produce base). Sodium hydroxide is often intentionally added to boiler water at noncorrosive levels. It may also be introduced unintentionally if chemical from a caustically regenerated demineralizer is inadvertently released into makeup water. Alkaline-producing salts may also contaminate the condensate by in-leakage through condensers, or from process streams. Poorly controlled or malfunctioning chemical-feed equipment may also cause excessive alkalinity.

The second contributing factor is the mechanism of concentration. Because sodium hydroxide and alkaline-producing salts are rarely present at corrosive levels in the bulk environment, a means of concentrating them must be present. Three basic concentration mechanisms exist:

1. Departure from nucleate boiling (DNB).   The term *nucleate boiling* refers to a condition in which discrete bubbles of steam nucleate at points on a metal surface. Normally, as these steam bubbles form, minute concentra-

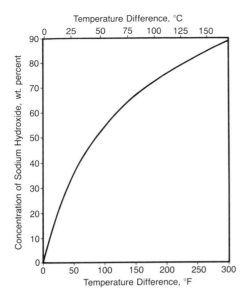

Figure 4.2 Sodium hydroxide content attainable in concentrating film of boiler water. (*Based on data from* International Critical Tables, *3:370(1928). Courtesy of Herbert H. Uhlig,* The Corrosion Handbook, *John Wiley and Sons, New York, 1948.)*

tions of boiler-water solids will develop at the metal surface, usually at the interface of the bubble and the water. As the bubble separates from the metal surface, the water will redissolve soluble solids such as sodium hydroxide (Fig. 1.1).

At the onset of DNB, the rate of bubble formation exceeds the rinsing rate. Under these conditions, sodium hydroxide, as well as other dissolved solids or suspended solids, will begin to concentrate (Fig. 1.3 and Fig. 4.2). The presence of concentrated sodium hydroxide and other concentrated corrosives will compromise the thin film of magnetic iron oxide, causing metal loss.

Under the conditions of fully developed DNB, a stable film or blanket of steam will form. Corrosives then concentrate at the edges of this blanket, causing metal loss at the perimeter. The metal at the interior of the blanket is left relatively intact.

2. Deposition.   A similar situation occurs when deposits shield the metal from the bulk water. Steam that forms under these thermally insulating deposits escapes and leaves behind a corrosive residue that can deeply gouge the metal surface (Fig. 4.3).

3. Evaporation at a waterline.   Where a waterline exists, corrosives may concentrate by evaporation, resulting in gouging along the waterline. In horizontal or slanted tubes, a pair of parallel longitudinal trenches may form (Fig. 4.4). If the tube is nearly full, the parallel trenches will coalesce into a single longitudinal gouge along the top of the tube (Fig. 4.5). In

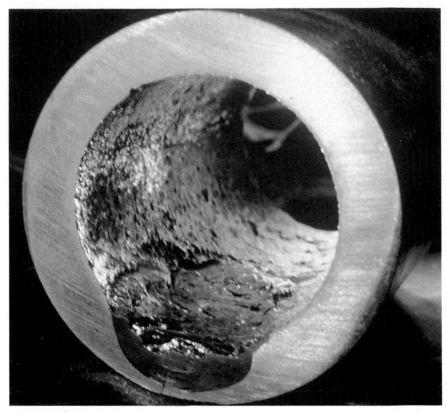

Figure 4.3 Deep caustic gouging beneath insulating internal deposits. *(Courtesy of National Association of Corrosion Engineers.)*

vertically oriented tubes, corrosive concentration at a waterline will yield a circumferential gouge.

Identification

If affected surfaces are accessible, caustic corrosion can be identified by simple visual examination. If not, nondestructive testing techniques such as ultrasonic testing may be required. Steam studies using a hydrogen analyzer may also be used to identify caustic corrosion.

Elimination

When the availability of sodium hydroxide or alkaline-producing salts and the mechanism of concentration exist simultaneously, they govern suscep-

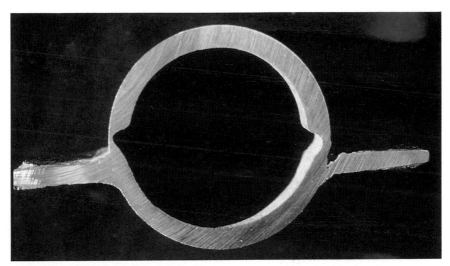

Figure 4.4 Caustic gouging along a longitudinal waterline. *(Courtesy of National Association of Corrosion Engineers.)*

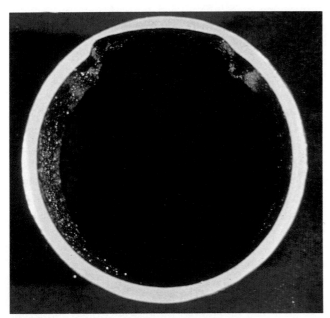

Figure 4.5 Caustic gouging resulting from evaporation at a waterline riding along the crown of the tube. *(Courtesy of National Association of Corrosion Engineers.)*

tibility to caustic corrosion. The following remedies may eliminate corrosion that depends on the availability of sodium hydroxide or alkaline-producing salts:

■ **Reduce the amount of available free sodium hydroxide.** This is the underlying concept that serves as the basis for coordinated phosphate programs implemented in high-pressure boilers.

■ **Prevent inadvertent release of caustic regeneration chemicals from makeup-water demineralizers.**

■ **Prevent in-leakage of alkaline-producing salts into condensers.** Because of the powerful concentration mechanisms that may operate in a boiler, in-leakage of only a few parts per million of contaminant may be sufficient to cause localized corrosion.

■ **Prevent contamination of steam and condensate by process streams.**

Although these remedies may eliminate corrosion that depends on the availability of sodium hydroxide or alkaline-producing salts, preventing localized concentration is the most effective means of avoiding caustic corrosion; it is also the most difficult to achieve. The methods for preventing localized concentration include:

■ **Prevent DNB.** This usually requires the elimination of hot spots, achieved by controlling the boiler's operating parameters. Hot spots are caused by excessive overfiring or underfiring, misadjusted burners, change of fuel, gas channeling, excessive blowdown, etc.

■ **Prevent excessive water-side deposition.** Tube sampling on a periodic basis (usually annually) may be performed to measure the relative thickness and amount of deposit buildup on tubes. Tube-sampling practices are outlined in ASTM D887-82. Consult boiler manufacturers' recommendations for acid cleaning.

■ **Prevent the creation of waterlines in tubes.** Slanted and horizontal tubes are especially susceptible to the formation of waterlines. Boiler operation at excessively low water levels, or excessive blowdown rates, may create waterlines. Waterlines may also be created by excessive load reduction when pressure remains constant. In this situation, water velocity in the boiler tubes is reduced to a fraction of its full-load value. If velocity becomes low enough, steam/water stratification occurs, creating stable or metastable waterlines.

Cautions

It is very difficult to distinguish localized attack by high-pH substances from localized attack by low-pH substances simply by visual examination. A formal metallographic examination may be required. Evaluating the types of concentrateable corrosives that may be contaminating boiler water will aid in the determination.

Because corrosion products may fill the depressions caused by caustic corrosion, the extent and depth of the affected area, and even the existence of a corrosion site, may be overlooked. Probing a suspect area with a hard, pointed instrument may aid in the determination, but because the corrosion products are often very hard, a corrosion site may remain undetected. The presence of sparkling crystalline magnetite does not necessarily indicate that caustic corrosion has occurred.

Related Problems

See also Chap. 1, "Water-Formed and Steam-Formed Deposits," and Chap. 6, "Low-pH Corrosion during Service."

CASE HISTORY 4.1

Industry:	Utility
Specimen Location:	Back wall of power boiler
Specimen Orientation:	Vertical
Years in Service:	6
Water-Treatment Program:	Coordinated phosphate
Drum Pressure:	1500 psi (10.3 MPa)
Tube Specifications:	2¾ in. (7.0 cm) outer diameter

Numerous caustic attacks on the back wall of a cyclone-fired boiler (Fig. 4.6) were all observed within a month. This type of attack had occurred once previously. This boiler was acid-cleaned every 18 to 24 months. The gouge was noted 1 year after the last acid cleaning.

Visual examinations disclosed hard layers of black, crystalline corrosion products covering the attack site. Measurement revealed a 42% reduction in tube-wall thickness. Microstructural examinations disclosed moderate overheating in the gouged region. Evidence revealed that DNB, rather than deposits, was responsible for caustic concentration in this case. Overfiring during start-up and low flow rates of the feedwater were suspected.

Figure 4.6 Region of caustic gouging along internal surface.

CASE HISTORY **4.2**

Industry:	Utility
Specimen Location:	Camera port, waterwall
Specimen Orientation:	Vertical and slanted, S-shaped
Years in Service:	25
Water-Treatment Program:	Coordinated phosphate
Drum Pressure:	2000 psi (13.8 MPa)
Tube Specifications:	3 in. (7.6 cm) outer diameter
Fuel:	Ground coal

Visual examinations disclosed a thickened patch of hard corrosion products adjacent to one bend (Fig. 4.7). Perforation of the wall had not occurred, but transverse cross sections cut through the site revealed substantial metal loss (Fig. 4.8).

The gouging was caused by sodium hydroxide that concentrated to corrosive levels along the site of a stable steam blanket, or perhaps at an isolated site of thermally insulating deposits. Previous failures of this type had not occurred in this region of the boiler. The boiler had been cleaned 4 years previously with a chelant and was in peaking service. Closer control over the water-treatment program might help prevent this type of problem in the future.

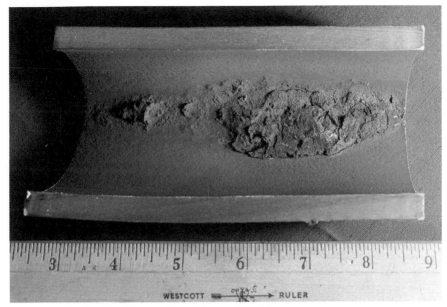

Figure 4.7 Patch of hard iron oxides on internal surface.

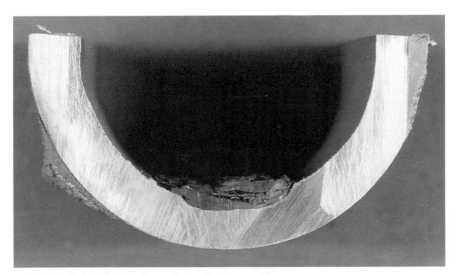

Figure 4.8 Cratered region beneath patch of iron oxides.

CASE HISTORY 4.3

Industry:	Utility
Specimen Location:	Bottom slag-screen tube
Specimen Orientation:	15° slope
Water-Treatment Program:	Coordinated phosphate
Drum Pressure:	2200 psi (15.2 MPa)
Tube Specifications:	3 in. (7.6 cm) outer diameter

A growing number of small leaks were occurring in lower slag-screen tubes of this boiler. One of the leaking tubes was removed for examination.

Figure 4.9 illustrates the appearance of the internal surface in the area of leakage. A small perforation of this rifled tube was observed in the center of a large, elliptical area of metal loss (Fig. 4.10). This area has a smooth, rolling metal-surface contour covered with a thick, irregular mound of coarsely stratified iron oxides. The rest of the internal surface had suffered no metal loss.

Since deposits were not present and evidence of a waterline was not observed, it can be assumed that concentration of the caustic material was caused by highly localized nonnucleate boiling (DNB). The rifling of the internal surface is designed to induce swirling of the water to prevent nonnucleate boiling and steam/water phase stratification. It is surprising, therefore, to find severe caustic gouging in this tube design.

However, this boiler was idle on weekends. It is possible that highly localized nonnucleate boiling may have occurred during start-up, before normal boiler-water circulation was fully established.

Figure 4.9 Thick, irregular mound of hard iron oxides covering perforation.

Figure 4.10 Perforation at bottom of crater.

CASE HISTORY 4.4

Industry:	Chemical process industry, ammonia plant
Boiler Type:	Heat-recovery boiler
Specimen Location:	Bottom of U-tube bundle
Specimen Orientation:	Horizontal on bottom, curving to vertical
Years in Service:	8
Water-Treatment Program:	Coordinated phosphate
Drum Pressure:	1500 psi (10.3 MPa)
Tube Specifications:	¾ in. (1.9 cm) outer diameter
Heat Source:	Reformer gas

A massive longitudinal perforation was evident on this tube section. The failure resulted from wall thinning on the internal surface. Metal loss was localized to the top of the tube and the area of metal loss was approximately 19 in. (48.3 cm) long (Fig. 4.11).

Examination revealed a distinct groove along the top of the internal surface; this groove diminished gradually in both depth and width as the tube assumed a vertical orientation. Vertical surfaces were not corroded. The groove surface was very smooth, slightly rolling in contour, and covered with a uniform coating of black iron oxide. The surface at the perimeter of the groove was marked by a dense population of deep, hemispherical pits that

Figure 4.11 Groove approximately 14 in. (35.6 cm) from perforation.

existed in a distinct band along the sides and at each end of the groove (Fig. 4.12).

The grooving resulted from the concentration of sodium hydroxide due to steam accumulation and channeling along the upper horizontal and slanted regions of the internal surfaces. In addition to the important contribution of the tube's horizontal and slanted orientation, the accumulation of steam resulted from a condition of either excessive heat input in this region or impaired coolant flow through these tubes. This problem might be corrected by reducing heat input in this part of the boiler or by increasing water velocity through these tubes.

These types of failures have also been prevented through the use of rifled tubes. Closer control of the chemical program for the boiler water also might eliminate this type of failure.

Figure 4.12 Band of hemispherical pits adjacent to groove. (Magnification: 6.5×.)

Figure 4.13   Grooved window section cut from inner bend.

CASE HISTORY 4.5

Industry:	Sugar
Specimen Location:	Top of riser tube near entrance Into steam drum
Specimen Orientation:	Slanted
Water-Treatment Program:	Polymer
Drum Pressure:	450 psi (3.1 MPa)
Tube Specifications:	3 in. (7.6 cm) outer diameter
Fuel:	No. 6 fuel oil

Visual examinations revealed a longitudinal groove on the internal surface along the top of the tube (Fig. 4.13). Perforation had not occurred, but as much as 60% of the tube wall had been corroded.

The entire top side of the internal surface exhibited shallow metal loss in a distinct band (Fig. 4.14), which narrowed and ended in a "spear point" near the end of the tube in the steam drum. Sparkling black crystals of magnetite were present in and around the groove. A total of six adjacent tubes had been similarly affected.

Steam channeling along the top of the slanted section led to concentration of sodium hydroxide. Steam channeling may indicate localized or general excessive heat input. If appropriate alteration of operating parameters does not eliminate the problem, the use of rifled tubes may be effective.

The affected boiler is operated continuously during 100 to 150 days of campaign operation twice per year. The boiler is not operated during the intervening period.

Figure 4.14 Internal surface showing wasted metal.

Chelant Corrosion

Locations

The steam drum is the area of the boiler most frequently affected by chelant corrosion. Steam-separation equipment, especially equipment designed to separate steam by centrifugal means, is most susceptible to this type of corrosion. Chelant corrosion may also occur in feedwater distribution lines, in the economizer, on the ends of downcomer tubes, and in regions of high heat flux in water-cooled tubes. Copper and copper-alloy impellers of feedwater pumps may also corrode if exposed to chelating agents.

General Description

Concentrated chelants may attack magnetite according to the following reaction:

$$Fe_3O_4 + Fe + 8H^+ + 4 \text{ chelant} \longrightarrow 4Fe(II) \text{ chelant} + 4H_2O$$

Surfaces attacked by chelants are typically very smooth and featureless. In situations where there is sufficient fluid velocity, the surfaces may have a smoothly rolling contour marked by "comet tails" and horseshoe-shaped depressions (Fig. 5.1). These features are aligned with the direction of flow. The metal surface will be uniformly covered with a submicroscopic film of

Figure 5.1 Comet-tail and horseshoe-shaped depressions.

either dull or glassy black material. A surface under active attack will be free of deposits and corrosion products; they will already have been chelated.

Occasionally, when excessively high levels of oxygen are present, the appearance of metal loss is altered. The surface under attack retains sharply defined islands of intact metal surrounded by a smooth plain of

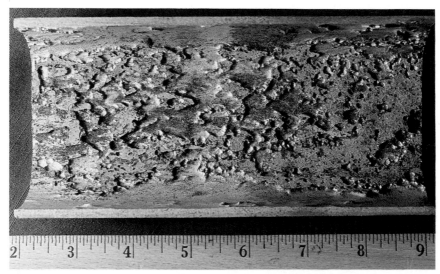

Figure 5.2 Chelant corrosion altered by the presence of excessive oxygen.

metal loss (Fig. 5.2). Some observers have reported that attack by chelants has given a jagged roughness, similar to that resulting from attack by a strong acid.

Critical Factors

Chelant corrosion can occur under a variety of circumstances. Overfeed of chelant is a frequently cited cause of chelant corrosion. However, at recommended residual levels of chelant, corrosion is possible in regions where a concentration mechanism is operating. The principal, and perhaps only, mechanism by which chelants are concentrated is by evaporation. Hence, chelant corrosion may occur where departure from nucleate boiling (DNB) occurs (Fig. 5.3).

Chelant corrosion of separation equipment in the steam drum, which does not contain heat-transfer surfaces, is apparently related to high fluid velocities. This type of corrosion is especially sensitive to the erosive effects of high-velocity fluids, and is sometimes encountered where these exist, even in the absence of heat transfer. Excessive levels of dissolved oxygen seem to act synergistically with excessive levels of chelant to produce jagged metal-surface contours.

Identification

Chelant corrosion can be visually identified if affected equipment is accessible. When access is very limited, as is generally the case with tubes,

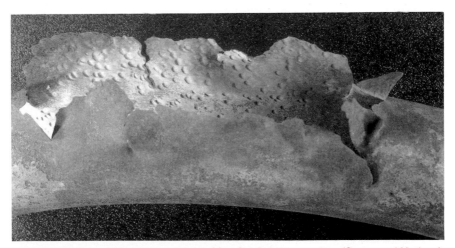

Figure 5.3 Tube rupture after severe metal loss by chelant corrosion. *(Courtesy of National Association of Corrosion Engineers.)*

nondestructive testing techniques such as ultrasonic testing may be used for identification.

Steam drum internals are particularly susceptible to chelant corrosion. Severe cases of corrosion have reduced cyclone separator cans to lacelike remnants.

Elimination

Close control of chelant and dissolved-oxygen levels is imperative. In order to eliminate chelant corrosion, certain actions can be taken.

First, special care must be exercised under conditions of inconsistent feedwater quality and when chelants are used in a dirty boiler. Second, elimination of hot spots in the firebox will prevent chelant corrosion of water-cooled tubes. Hot spots are frequently caused by improper boiler operation, improper boiler maintenance, and design deficiencies. Excessive overfiring or underfiring, misadjusted burners, change of fuel, gas channeling, excessive blowdown, and dislodged refractory have all been linked to the creation of hot spots as well. Many cases of chelant corrosion can be reduced in severity or eliminated altogether by reducing fluid velocity and eliminating turbulent flow.

Cautions

Chelant corrosion can give a smooth, featureless contour to metal surfaces. This feature, coupled with the fact that chelant-corroded metals typically have a black, passive appearance, masks the fact that active corrosion is occurring. In these cases, thickness measurements of suspect areas may reveal a problem.

Simple erosion by high-velocity steam or water may also yield an appearance that closely resembles that of corrosion by chelants.

Every effort should be made to avoid exposing copper-based alloys to chelants.

Related Problems

See also Chap. 7, "Low-pH Corrosion during Acid Cleaning," and Chap. 17, "Erosion."

CASE HISTORY 5.1

Industry:	Steel
Specimen Location:	Wall tube at bottom waterwall header
Specimen Orientation:	Vertical
Years in Service:	14
Water-Treatment Program:	Chelant
Drum Pressure:	900 psi (6.2 MPa)
Tube Specification:	2½ in. (6.3 cm) outer diameter
Fuel:	No. 6 oil and various waste gases

Failure of a section adjacent to the tube, illustrated in Fig. 5.4, occurred during hydrotesting. Severe internal wastage was confined to 6 to 9 in. (15.2 to 22.9 cm) of tube length in each of four adjacent tubes. Metal loss within the affected region varied. In areas of severe metal loss the appearance was smooth and wavelike. Discrete, isolated pits were evident in areas of moderate metal loss.

The highly localized attack on this section illustrates the importance of velocity and turbulence in an erosion-corrosion process involving chelants. Metal loss is severe in regions of turbulence at the inlet end of the tube from the header. The attack moderates with distance along the tube, until it ceases altogether where laminar flow is established. The chelant alone was not sufficiently aggressive to cause corrosion. Rather, metal loss resulted from the synergistic interaction of the chelant with localized turbulence.

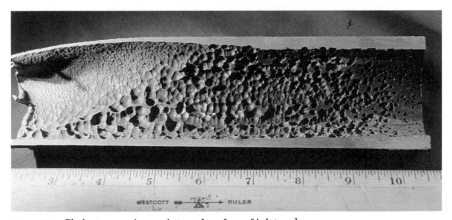

Figure 5.4   Chelant corrosion on internal surface of inlet end.

CASE HISTORY 5.2

Industry:	Steel
Specimen Location:	Wall tube
Specimen Orientation:	Vertical bend
Years in Service:	25
Water-Treatment Program:	Chelant
Drum Pressure:	800 psi (5.5 MPa)
Tube Specifications:	3 in. (7.6 cm) outer diameter
Fuel:	Blast-furnace gas

The massive rupture of the wall tube (Fig. 5.3) was the first to occur in the boiler. The edges of the rupture are very thin, and a population of horseshoe-shaped depressions oriented in the direction of flow can be seen along the wasted internal surface. Microstructural examinations revealed that overheating had not occurred.

The rupture occurred on the cold side of the tube. The insulation installed to protect the cold side of this tube apparently had been dislodged by steam impingement from a leaking superheater tube. Hot furnace gases that reached the unprotected back side of the tube caused unstable film boiling and concentration of chelant from the boiler water.

The combination of concentrated chelant and fluid velocity caused the thinning of the internal surface apparent in Fig. 5.3. Thinning of the tube wall in this manner continued until stresses imposed by normal internal pressure exceeded the tensile strength of the thinned tube wall.

CASE HISTORY 5.3

Industry:	Chemical process
Specimen Location:	Feedwater-pump impeller
Specimen Orientation:	Vertical
Years in Service:	3
Water-Treatment Program:	Chelant
Tube Specifications:	8 in. (20.3 cm) outer diameter
Tube material:	Bronze

This impeller, illustrated in Fig. 5.5, was removed from the fifth stage of a feedwater pump. Broad areas of general wastage are apparent. Areas of metal loss are confined principally to regions of high turbulence such as the vane edges and discharge throat. Horseshoe-shaped depressions are visible in some regions.

Figure 5.5 Chelant corrosion on feedwater pump.

Close examination of wasted surfaces under a low-power stereoscopic microscope revealed dendrites and other microstructural features of the casting. Chelants are aggressive toward copper and copper-based alloys and should be fed well downstream of copper-alloy impellers.

CASE HISTORY 5.4

Industry:	Brewing
Specimen Location:	Steam-drum end of downcomer tubes
Specimen Orientation:	10 to 15° from vertical
Years in Service:	35
Water-Treatment Program:	Chelant
Drum Pressure:	485 psi (3.3 MPa)
Tube Specifications:	2½ in. (6.3 cm) outer diameter

Figure 5.6 illustrates the appearance of the internal surfaces of 12 adjacent downcomer tubes that have suffered severe internal corrosion. Although the same water-treatment program had been utilized for this boiler for 17 years, this corrosion occurred within a period of 7 months.

The affected tubes were located at one end of the boiler, within 9 in. (22.9 cm) of the feedwater line in the steam drum. It is surmised that feedwater was short-circuiting through the affected area.

Oxygen pitting was present along the waterline throughout the steam

Figure 5.6 Internal surface of downcomer tube.

drum, and on the steam-separation canisters. This is evidence that excessive levels of dissolved oxygen were present in the feedwater. The excessive oxygen levels, coupled with the presence of the chelant, resulted in the corrosion observed. The metal loss produced smooth, rolling, wavelike surface contours and islands of intact metal in affected areas (Fig. 5.7).

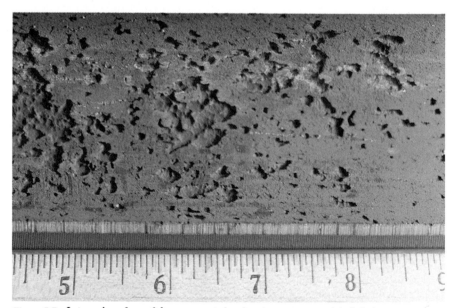

Figure 5.7 Internal surface of downcomer tube showing oxygen-assisted chelant corrosion.

The entire internal surface was lightly covered with deposits and iron oxides. The corroded metal was covered with a shiny black film under these deposits and iron oxides. The layer of iron oxides and deposits covering corroded surfaces reveals that corrosion had not been active recently, and therefore was not continuous, but intermittent.

This type of failure can be prevented by gaining control of dissolved oxygen. Preventing short-circuiting of the feedwater is also a solution to this type of corrosion.

Low-pH Corrosion during Service

Locations

Generally, in-service acid corrosion is confined to water-cooled tubes in regions of high heat flux; slanted or horizontal tubes; locations beneath heavy deposits; and heat-transfer regions at or adjacent to backing rings at welds, or to other devices that disrupt flow.

General Description

Although relatively rare, a general depression of bulk-water pH may occur if certain contaminants gain access to the boiler. Boilers using water of low buffering capacity can realize a bulk pH drop to less than 5 if contaminated with seawater, hydrochloric acid, or sulfuric acid.

The concern of this chapter, however, is with the more common creation of localized pH conditions. Two circumstances must exist simultaneously to produce this condition. First, the boiler must be operated outside of normal, recommended water-chemistry parameters. This may happen if condenser in-leakage occurs when seawater, or water from a recirculating water system using cooling towers, is used. Another source of contamination is the inadvertent release of acidic regeneration chemicals from a makeup-water demineralizer into the feedwater system.

The second condition that must exist to produce low-pH conditions is a mechanism for concentrating acid-producing salts. This condition exists where boiling occurs and adequate mixing is hindered by the presence of porous deposits or crevices. Where deposits or crevices are present, a concentration of acid-producing salts may induce hydrolysis to produce localized low-pH conditions, while the bulk water remains alkaline.

$$M^+Cl^- + H_2O \rightarrow MOH\downarrow + H^+Cl^-$$

Wherever low-pH conditions exist, the thin film of magnetic iron oxide is dissolved and the metal is attacked. The result is gross metal loss. This loss may have smooth, rolling contours similar in appearance to caustic gouging. The gouged area will frequently be covered with nonprotective iron oxides.

Critical Factors

Two critical factors contribute to low pH. The first is the availability of free acid or acid-producing salts (i.e., salts whose solution with water may produce acid). Unintentional additions of free acid may arise from inadvertent release of acid regeneration chemicals from a makeup-water demineralizer into the makeup water. Acid-producing salts may contaminate the condensate by in-leakage through condensers, or from process streams. Because of the powerful concentration mechanisms that may operate in a boiler, in-leakage of only a few parts per million of contaminant may be sufficient to cause localized low-pH corrosion in unbuffered boiler water. Residual contamination from chemical cleaning and poorly controlled or malfunctioning feedwater-chemical equipment may also foster localized low-pH conditions.

The second contributing factor is the mechanism of concentration. Since free acid or acid-producing salts are not usually present at corrosive levels in the bulk environment, some means of concentrating them must be present. Three basic concentration mechanisms exist:

1. Departure from nucleate boiling (DNB). *Nucleate boiling* refers to a condition in which discrete bubbles of steam nucleate at points on a metal surface. Normally, as these steam bubbles form, minute concentrations of boiler-water solids will develop at the metal surface, usually where the bubble and water interface. As the bubble separates from the metal surface, the water will redissolve soluble solids (Fig. 1.1).

At the onset of DNB, the rate of bubble formation exceeds the rinsing rate. Dissolved solids, or suspended solids, will begin to concentrate (Fig. 1.3). The presence of concentrated corrosives will compromise the thin film of magnetic iron oxide, causing metal loss.

Under the conditions of fully developed DNB, a stable film or blanket of steam will form. Corrosives then concentrate at the edges of this blanket, causing metal loss at the perimeter. The metal at the interior of the blanket is left relatively intact.

2. Deposition. A similar situation occurs when deposits shield the metal from the bulk water. Steam that forms under these thermally insulating deposits escapes and leaves behind a corrosive residue that can deeply gouge the metal surface.

3. Evaporation at a waterline. Where a waterline exists, corrosives may concentrate by evaporation, resulting in gouging along the waterline. In horizontal or slanted tubes, a pair of parallel, longitudinal trenches may form. If the tube is nearly full, the parallel trenches will coalesce into a single longitudinal gouge along the top of the tube. In vertically oriented tubes, corrosive concentration at a waterline will yield a circumferential gouge.

Identification

Simple visual examination is sufficient if affected surfaces are accessible. If not, nondestructive testing techniques, such as ultrasonic testing, may be required. Steam studies using a hydrogen analyzer may also identify localized low-pH corrosion.

Elimination

When the availability of free acid or acid-producing salts and the mechanism of concentration exist simultaneously, they govern susceptibility to localized low-pH corrosion. The following remedies may eliminate low-pH corrosion based on the availability of free acids or acid-producing salts:

■ **Prevent inadvertent release of acidic regeneration chemicals from makeup-water demineralizers.**

■ **Prevent in-leakage of acid-producing salts, such as calcium chloride and magnesium chloride, into condensers.** Because of the powerful concentration mechanisms that may operate in a boiler, in-leakage of only a few parts per million of contaminant may be sufficient to cause localized corrosion.

- **Prevent contamination of steam and condensate by process streams.**

Although these remedies may eliminate corrosion based on the availability of free acids or acid-producing salts, preventing localized concentration is the most effective means of avoiding low-pH corrosion. It is also the most difficult to achieve. The methods for preventing localized concentration include:

- **Prevent DNB.**   This usually requires the elimination of hot spots, accomplished by controlling the boiler's operating parameters. Hot spots are caused by such things as excessive overfiring or underfiring, misadjusted burners, change of fuel, gas channeling, and excessive blowdown.

- **Prevent excessive water-side deposition.**   Tube sampling on a periodic basis (usually annually) may be performed to measure the relative thickness and amounts of deposit buildup on tubes. Tube-sampling practices are outlined in ASTM D887-82. Consult boiler manufacturers' recommendations for acid cleaning.

- **Prevent the creation of waterlines in tubes.**   Slanted and horizontal tubes are especially susceptible to the formation of waterlines. Operation of the boiler at excessively low water levels, or excessive blowdown rates, may create waterlines. Waterlines may also be created by excessive load reduction when pressure remains constant. Under this circumstance, the water velocity in the boiler tubes is reduced to a fraction of its full load value. If it becomes low enough, steam/water stratification occurs, creating stable or metastable waterlines.

Cautions

It is very difficult to distinguish localized attack by low-pH substances from localized attack by high-pH substances simply by visual examination. Distinguishing between the two may require a formal metallographic examination. Evaluating the types of concentrateable corrosives that may be contaminating the boiler water will aid in the determination.

Because corrosion products may fill the depressions caused by low-pH corrosion, the extent and depth of the affected area, and even the existence of a corrosion site, may be overlooked. Probing a suspect area with a hard, pointed instrument may aid in the determination, but because the corrosion products are often hard, a corrosion site may remain undetected.

Related Problems

See also Chap. 4, "Caustic Corrosion," and Chap. 14, "Hydrogen Damage."

CASE HISTORY 6.1

Industry:	Utility
Specimen Location:	Side wall
Specimen Orientation:	Vertical
Years in Service:	30
Water-Treatment Program:	Coordinated phosphate
Drum Pressure:	2000 psi (13.8 MPa)
Tube Specifications:	3 in. (7.6 cm) outer diameter

Numerous occurrences of the type of gouging illustrated in Fig. 6.1 had resulted in an extensive tube-replacement program. Most replacements were in one of the boiler's side walls. This type of corrosion was a recurring

Figure 6.1 Low-pH gouging.

Figure 6.2 Deep gouge on internal surface. (Magnification: 7.5X.)

problem, but the frequency had recently increased. The boiler was in peaking service.

Figure 6.2 illustrates the extent and appearance of the gouging. The crater was filled with thick, hard, iron oxides and elemental copper. (Most gouging had occurred downstream of circumferential welds.)

Analysis of corrosion products on the internal surface of the crater revealed significant amounts of chloride.

The large percentage of occurrences immediately downstream of circumferential welds indicates that disrupted water flow across the weld was instrumental in establishing a concentration site for corrosive substances.

CASE HISTORY 6.2

Industry:	Utility
Specimen Location:	Wall tube
Specimen Orientation:	Slightly off vertical
Years in Service:	26
Water-Treatment Program:	Coordinated phosphate
Drum Pressure:	1900 psi (13.1 MPa)
Tube Specifications:	3¼ in. (8.3 cm) outer diameter

Figures 6.3 and 6.4 illustrate the massive rupture that occurred in a distinct zone of deep metal loss along the internal surface of this tube. Surrounding areas of the internal surface were unaffected and quite smooth.

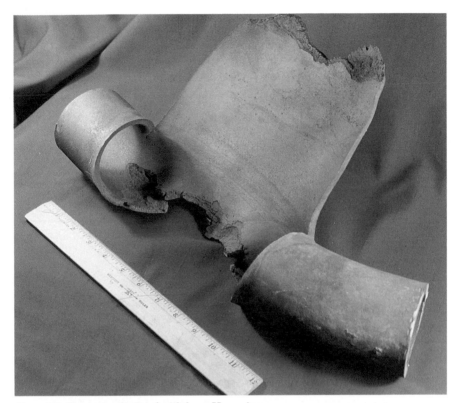

Figure 6.3   Rupture associated with low-pH gouging.

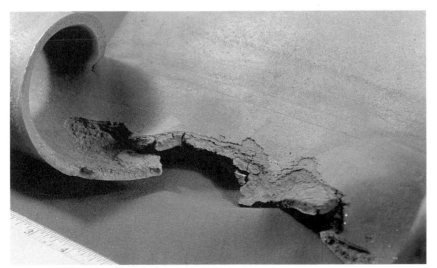

Figure 6.4   Metal loss at the origin of rupture.

Microstructural examinations revealed extensive hydrogen damage in the tube wall immediately below the gouged zone.

The visual and microstructural appearance of the gouged region was consistent with that of low-pH exposure. The hydrogen damage that is associated with the low-pH gouging indicates that corrosion occurred during boiler operation.

Concentration of low-pH corrosive may have occurred beneath deposits that were dislodged from the internal surface at the time of rupture. It is also possible that concentration occurred as a result of the presence of a steam blanket that resulted from nonnucleate boiling.

CASE HISTORY 6.3

Industry:	Utility
Specimen Location:	Waterwall
Specimen Orientation:	Slanted (nose arch)
Years in Service:	10
Water-Treatment Program:	Congruent control
Drum Pressure:	2700 psi (18.6 MPa)
Tube Specifications:	3 in. (7.6 cm) outer diameter

Figure 6.5 illustrates one of many tubes that had sustained similar severe corrosion. Grooves were located along the top (crown) of each tube. The convergence of the groove, apparent in Fig. 6.5, marked the position where the slanted tube assumed a vertical orientation. Metal loss was not observed downstream of this point.

Corroded areas were covered with black, powdery deposits (Fig. 6.6). These deposits covered a second layer of light-colored material that was

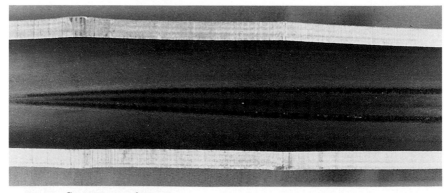

Figure 6.5 Convergence of groove.

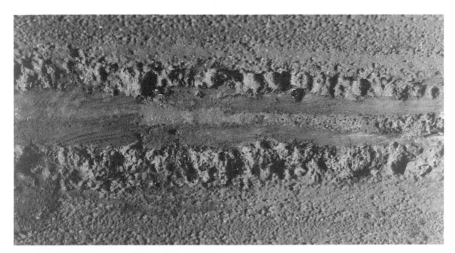

Figure 6.6   Appearance of deposits covering groove.

directly on the metal. The light-colored material was present only in corroded regions. On either side of the groove a hard coating of protective magnetite covered the smooth internal surface. The contour of the corroded surface following removal of the deposits and the light-colored material is shown in Fig. 6.7.

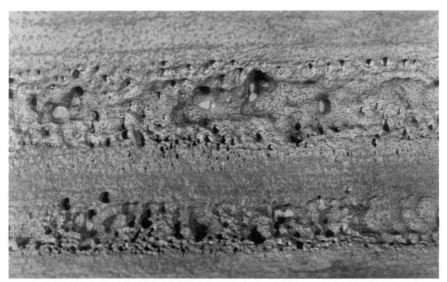

Figure 6.7   Contour of groove after deposit removal. *(Courtesy of Electric Power Research Institute.)*

Microstructural analyses revealed that the grooved tube wall had sustained mild overheating. Analyses of the substances covering corroded surfaces revealed that the light-colored material on the metal surface was highly crystalline and was composed of iron and phosphorus. X-ray diffraction studies of this material revealed iron phosphate and sodium iron phosphate.

Gravitational stratification of water and steam established relatively stable steam channels along the tops of the slanted tubes. Evaporative concentration of dissolved solids within this channel furnished the corrodent. Careful chemical analyses revealed that iron phosphate and sodium iron phosphate were confined to corroded metal surfaces.

While a diagnosis of caustic gouging might explain this corrosion, the microstructural contours of corroded surfaces, the absence of dense iron oxide or crystalline magnetite, and the specific water chemistry of the boiler in this case all suggest corrosion by a weak, phosphorus-containing acid.

CASE HISTORY 6.4

Industry:	Chemical process
Specimen Location:	Economizer
Specimen Orientation:	Horizontal
Years in Service:	7
Water-Treatment Program:	Chelant
Drum Pressure:	155 psi (1.1 MPa)
Tube Specifications:	2⅜ in. (6.0 cm) outer diameter

Corrosion was a recurrent problem in the economizer of this gas-fired boiler. Failures occurred only in the hot end of the economizer at the beginning of the finned area (Fig. 6.8). Heat fluxes in this area were 40% higher than design values.

Corrosion formed a large elliptically shaped gouge covered with a thick, irregular layer of hard, dark iron oxides. Microstructural examinations of the iron oxides covering this gouge revealed a laminated structure typical of low-pH gouging.

Nonnucleate boiling in this area of the economizer (caused by excessively high heat fluxes) furnished the means for concentrating corrosive, low-pH substances. The source of the acid-producing salts was undetermined, but may have been caused by leakage of sodium chloride from the regeneration of a water softener.

Figure 6.8 Gouging along internal surface opposite fins.

CASE HISTORY 6.5

Industry:	Chemical process
Specimen Location:	Tube from an ethylene-cracking furnace
Specimen Orientation:	Horizontal
Years in Service:	Unknown
Water-Treatment Program:	Antifoulant added
Drum Pressure:	175 psi (1.2 MPa)
Tube Specifications:	4½ in. (11.4 cm) outer diameter

The perforation illustrated in Fig. 6.9 occurred along the bottom of the first tube pass in the convection section of the furnace. Steam and ethane, in the ratio of 1 to 3, pass through the tube.

The metal loss, illustrated in Fig. 6.10, occurred on the internal surface of the tube and was responsible for the perforation. The corrosion produced a smooth, rolling surface contour.

Dark deposits, which gave a low-pH indication when wetted, surrounded the corrosion site. Analysis of these deposits and corrosion products revealed high levels of phosphorus and iron. It is possible that an acidic, phosphorus-containing substance had concentrated along the hot side of the tube, resulting in this deterioration.

Figure 6.9   Perforation resulting from corrosion of internal surface.

Figure 6.10   Appearance of metal loss on internal surface.

Low-pH Corrosion during Acid Cleaning

Locations

Corrosion of the internal surfaces of a boiler that results from low-pH exposure may occur during acid cleaning if proper procedures are not followed. One of the first areas to be affected is the tube ends inside the mud and steam drums. Hand-hole covers, drum manholes, and shell welds may also be affected. Heat-transfer surfaces (Fig. 7.1) and weldments (Fig. 7.2) may experience vigorous attack. Shielded regions within crevices, behind backing rings, and under remaining deposits may prevent proper neutralization of the cleaning acid. This results in vigorous localized attack of the metal once the boiler is returned to service. In general, any surface exposed to acid is susceptible (Fig. 7.3).

General Description

Attack of any metal surface of a boiler by strong acid is generally unmistakable. The surface usually has a rough or jagged appearance, depending on the severity of the attack (Figs. 7.1 and 7.4). A close examination will disclose discrete pits, which are frequently undercut. On boiler tubes, the pits will frequently be aligned longitudinally along the tube wall (Fig. 7.1).

Figure 7.1   Acid corrosion on the internal surface of a wall tube. *(Courtesy of Electric Power Research Institute.)*

Figure 7.2   Acid corrosion at weld.

Figure 7.3 Acid corrosion of attachment hardware.

Figure 7.4 Jaggedness associated with severe acid corrosion. *(Courtesy of Electric Power Research Institute.)*

Corrosion of steel by acids is a natural consequence of steel's thermodynamic instability in these environments. Steel will corrode spontaneously in most acids. During the corrosion reaction, iron displaces hydrogen from solution. That is, the iron is oxidized and iron ions go into solution. Hydrogen ions are reduced and form hydrogen bubbles at the metal surface.

$$Fe + 2H^+ + Cl^- \rightarrow H_2\uparrow + Fe^{2+} + Cl^-$$

To stifle this corrosion process, inhibitors are added to acid-cleaning solutions used in boilers.

Critical Factors

Uncontrolled acid corrosion of a boiler during cleaning generally results from an unanticipated and unintentional deviation from standard conditions or standard practice. Many deviations are possible and may include events such as thermally induced breakdown of the inhibitor, inappropriate selection of cleaning agent or cleaning strength, excessive exposure times, excessive exposure temperatures, and failure to neutralize completely.

Identification

Simple visual examinations are generally adequate to identify acid corrosion. Attack generally can first be observed at tube ends in mud and steam drums, at ends of sheet or plate steel, and at ends of bolts. All areas of the boiler may not be affected to the same degree. Stressed metal, welded joints, crevices, and other shielded regions may suffer more intense damage. Damage assessment in visually inaccessible areas may require either nondestructive testing techniques, such as ultrasonic testing, or removal of a tube section.

Elimination

Mitigation of low-pH corrosion of boiler equipment during acid cleaning requires close monitoring of the entire cleaning procedure. The following are merely examples of the various operational parameters that must be monitored and evaluated during the procedure.

Deposit-weight determinations.   Appropriate selection of several tubes for deposit-weight measurements will aid in determination of proper acid strength, exposure time, and total quantity of acid required to adequately clean the boiler.

Deposit analyses. Deposit analyses will help in determination of appropriate cleaning agents and the sequence in which the agents should be used.

Temperature of cleaning solution. Both the solution temperature and the metal temperature should be safely below the thermal breakdown point of the inhibitor.

Monitoring. Iron content, copper content, and cleaning-solution strength should be monitored at periodic intervals during the boiler cleaning. Chemistry of neutralizer should be monitored following the boiler's exposure to the acid.

Visual inspection. Inspection of tubes, mud drums, and steam drums should follow cleaning.

Cautions

Vigorous oxygen pitting and cavitation damage have been mistaken for attack by strong acid. Generally, proper distinction can be made by observing that attack by strong acid generally affects all exposed surfaces. Oxygen corrosion tends to occur in specific areas, such as the economizer, return bends in the superheater, or perhaps along the waterline in the steam drum. Cavitation damage also tends to be location-specific, and most commonly affects pump impellers. Certain forms of chelant corrosion may resemble acid corrosion, but again, chelant corrosion tends to be specific to steam drum internals.

Related Problems

See also Chap. 5, "Chelant Corrosion"; Chap. 8, "Oxygen Corrosion"; and Chap. 18, "Cavitation."

CASE HISTORY 7.1

Industry:	Utility
Specimen Location:	Wall tubes
Specimen Orientation:	Vertical
Years in Service:	30
Water-Treatment Program:	Phosphate and low-excess hydroxide
Drum Pressure:	875 psi (6.0 MPa)
Tube Specifications:	2½ in. (6.3 cm) outer diameter
Cleaning Solution:	Mineral acid

Following an acid cleaning of the boiler, the damage apparent in Figs. 7.5 and 7.6 was observed. Close visual examinations of these tubes revealed very fine, discontinuous fissures running longitudinally down the internal surfaces. These fissures, which appeared to coincide with a series of faint mandrel marks, were deeper at the end that had been rolled into the drum. The tube illustrated in Fig. 7.6 also shows deep transverse fissuring where it had been rolled into the drum. Examination of cross-sectional profiles of tube surfaces revealed they were jagged, undercut, and free of normally present iron oxides.

Attack by strong mineral acid is responsible for the delineation and deepening of mandrel marks running down the bore of the tube, as well as for the development of deep fissures near the end of the tubes that had been rolled into the drum. The delineation and deepening of the mandrel marks is associated with residual stresses in the metal at these sites from the tube-fabrication process. The presence of deep fissures of various orientations at the ends rolled into the drums is associated with residual stresses resulting from the tube-rolling process. Strong mineral acid characteristically attacks

Figure 7.5 Longitudinal fissures near rolled end.

Figure 7.6 Transverse fissures near rolled end.

stressed metal areas more aggressively than unstressed areas; this is because of the greater energy content of metal associated with residual stresses.

CASE HISTORY 7.2

Industry:	Utility
Specimen Location:	Wall tube
Specimen Orientation:	Vertical
Years in Service:	15
Water-Treatment Program:	Congruent control
Drum Pressure:	1960 psi (13.5 MPa)
Tube Specifications:	2½ in. (6.3 cm) outer diameter
Cleaning Solution:	Citric acid

Attack of metal surfaces by organic acids generally differs significantly from that produced by strong mineral acids. Given equivalent exposures, organic acids generally cause less corrosion. In addition, surface features of metals subject to excessive exposure to organic acids are typically less jagged and less undercut than those of metals subject to excessive exposure to strong mineral acids.

Attack by critic acid is illustrated in Fig. 7.7. Visually, the surface has an etched, bright metallic appearance. Close examination reveals irregular islands of uncorroded metal, which, in this case, stand 0.005 in. (0.12 mm) above the surrounding corroded surface. Essentially the entire internal

Figure 7.7   Appearance of attack by citric acid. (Magnification: 7.5X.)

surface, both hot and cold sides, was similarly affected. This condition is consistent with attack during acid cleaning.

CASE HISTORY 7.3

Industry:	Utility
Specimen Location:	Wall tube
Specimen Orientation:	Vertical
Years in Service:	30
Water-Treatment Program:	Congruent control
Drum Pressure:	1500 psi (10.3 MPa)
Tube Specifications:	3 in. (7.6 cm) outer diameter
Cleaning Solution:	Mineral acid

Figures 7.2 and 7.8 illustrate attack of the internal surfaces of wall tubes opposite external circumferential welds. Cross-sectional profiles of the internal surface in these attacked zones revealed undercutting and jaggedness. Residual stresses remaining from the weld make these sites subject to preferential corrosion when exposed to strong acid.

Figure 7.8 Acid corrosion at welds.

CASE HISTORY 7.4

Industry:	Utility
Specimen Location:	Side-wall tube
Specimen Orientation:	Vertical
Years in Service:	30
Water-Treatment Program:	Congruent phosphate
Control Drum Pressure:	1500 psi (10.3 MPa)
Tube Specifications:	3 in. (7.6 cm) outer diameter
Cleaning Solution:	Mineral acid

A number of fairly deep, overlapping pits were observed in a circumferential zone along the internal surface opposite circumferential welds on the external surfaces of the tubes (Fig. 7.9). Other areas of attack were noted in regions where welds were not present (Fig. 7.10). Attack in these areas was more pronounced along the tube seam.

Deposits overlying the attack sites reveal that the corrosion occurred in the past, probably during the last acid cleaning of the boiler.

Preferential attack of welds may be due to residual stresses associated with the weld, or to pores or crevices existing in the weld zone. Preferential attack of a tube seam may be due to segregated impurities in the seam, or to a crevice resulting from incomplete fusion of the seam.

Figure 7.9 Acid corrosion at weld.

Figure 7.10 Acid corrosion along internal surface. (Magnification: 7.5X.)

CASE HISTORY 7.5

Industry:	Pulp and paper
Specimen Location:	Belly plate
Specimen Orientation:	Horizontal
Years in Service:	16
Water-Treatment Program:	Polymer
Drum Pressure:	600 psi (4.1 MPa)
Cleaning Solution:	Mineral acid

Visual examinations of belly-plate surfaces revealed shallow patches of irregular metal loss on all top and bottom surfaces. The sites of metal loss tended to be aligned in parallel rows. Figure 7.11 illustrates the appearance of the metal loss, although a thin coating of iron oxide tends to soften the normally sharp edges.

This case illustrates the importance of supporting observations in establishing low-pH corrosion as the cause of metal loss. Figure 7.11 lacks the classical, jagged, undercut surface profile normally associated with attack by a strong acid. However, the following additional observations strongly support the diagnosis:

- Alignment of attack sites in parallel rows oriented in the rolling direction of the steel from which the plate was fabricated
- Preferential attack of welds and tube ends in the steam drum
- Jagged, undercut, cross-sectional profiles of belly-plate surfaces apparent in metallographic examinations

Figure 7.11   Metal loss from strong acid. (Magnification: 7.5X.)

CASE HISTORY 7.6

Industry:	Pulp and paper
Specimen Location:	Screen tube
Specimen Orientation:	Vertical
Years in Service:	13
Water-Treatment Program:	Coordinated phosphate
Drum Pressure:	1250 psi (8.6 MPa)
Tube Specifications:	2¼ in. (5.7 cm) outer diameter
Cleaning Solution:	Mineral acid

A combination of welding defects in circumferential welds, and attack of the weld area by strong mineral acid over a series of five acid cleanings during a 13-year period, resulted in tube perforations. Perforations at the weld sites led to secondary failures of adjacent tubes by erosion caused by steam escaping at high velocity.

Close examination of the internal surfaces of the welds revealed highly localized areas of corrosion, which produced jagged, undercut surface contours typical of attack by strong mineral acid (Fig. 7.12).

Open crevices, pores, and pits in weld areas are sites where acid may survive neutralization steps following acid cleaning. Acid corrosion can enlarge and deepen these sites, resulting in eventual perforation of the tube wall at the weld.

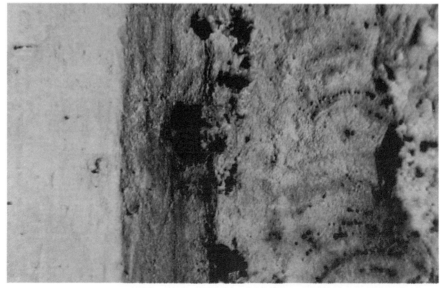

Figure 7.12   Acid attack along weld on internal surface. (Magnification: 7.5x.)

CASE HISTORY 7.7

Industry:	Textile
Specimen Location:	Fire tube
Specimen Orientation:	Horizontal
Water-Treatment Program:	Phosphate
Drum Pressure:	100 psi (0.6895 MPa)
Tube Specifications:	2½ in. (6.3 cm) outer diameter
Cleaning Solution:	Mineral acid

Hydrotesting following acid cleaning of the boiler revealed several leaking tubes. Examination of one of the leaking tubes revealed profuse pitting (Figs. 7.13 and 7.14). External surfaces were covered with a film of brown iron oxides.

Although some pitting may have occurred as a result of excessively high oxygen levels in the boiler water during idle times, the primary mode of attack was insufficiently controlled exposure to strong mineral acid during acid cleaning. Note the large population of pit sites, the general deterioration, and the appearance of fine longitudinal fissures characteristic of attack by a strong acid (Fig. 7.13).

Figure 7.13 Acid corrosion on the external surface of a fire tube.

Figure 7.14   Close-up of external surface shown in Fig. 7.13.

Oxygen Corrosion

Locations

Although relatively uncommon in an operating boiler, oxygen attack is a problem frequently found in idle boilers. The entire boiler system is susceptible, but the most common attack site is in superheater tubes. Reheater tubes are also susceptible, especially where moisture can collect in bends and sags in the tubes.

In an operating boiler the first areas to be affected are the economizer and the feedwater heaters. In cases of severe oxygen contamination, metal surfaces in other areas of the boiler may be affected — for example, surfaces along the waterline in the steam drum, and in the steam-separation equipment. In all cases, considerable damage can occur even if the period of oxygen contamination is short.

General Description

One of the most frequently encountered corrosion problems results from exposure of boiler metal to dissolved oxygen. Since the oxides of iron are iron's natural, stable state, steels will spontaneously revert to this form if conditions are thermodynamically favorable. Generally, conditions are favorable if steel that is not covered by the protective form of iron oxide (magnetite) is exposed to water containing oxygen. The following reaction occurs:

$$2Fe + H_2O + O_2 \rightarrow Fe_2O_3 + 2H\uparrow$$

This reaction is the basis for the intensive mechanical and chemical deaeration practices that are typical of sound water-treatment programs. These practices are generally successful. In fact, occurrences of oxygen corrosion in boilers are generally confined to idle periods.

For example, moisture condensing on the walls of an idle superheater tube will dissolve atmospheric oxygen. Fractures in the protective magnetite are caused by contraction stresses as the superheater is cooled to ambient temperatures. The fracture sites furnish anodic regions where oxygen-containing moisture can react with bare, unprotected metal. The result may be deep, distinct, almost hemispherical pits (Fig. 8.1), which may be covered at times with caps of corrosion products (Fig. 8.2). Frequently, pitting will occur at the bottom of U-shaped superheater pendants where moisture can accumulate (Fig. 8.3).

In addition to tube-wall perforation, oxygen corrosion is troublesome from another perspective. Oxygen pits can act as stress-concentration sites, thereby fostering the development of corrosion-fatigue cracks, caustic cracks, and other stress-related failures.

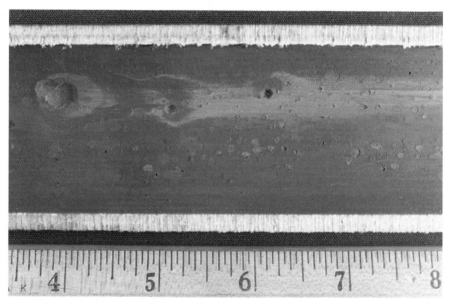

Figure 8.1   Oxygen pits in section of superheater tube. *(Courtesy of Electric Power Research Institute.)*

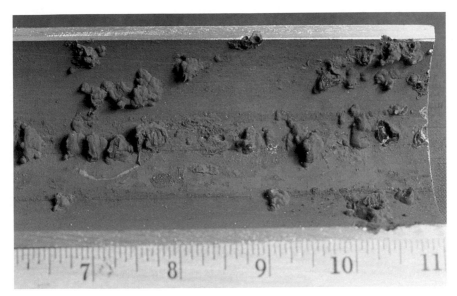

Figure 8.2   Caps of iron oxide covering pit sites.

Figure 8.3   Oxygen pits along bottom of superheater pendant. *(Courtesy of National Association of Corrosion Engineers.)*

Critical Factors

The three critical factors governing the onset and progress of oxygen corrosion include the presence of moisture or water, the presence of dissolved oxygen, and an unprotected metal surface.

The corrosiveness of water increases as temperature and dissolved solids increase, and as pH decreases. Aggressiveness generally increases with an increase in oxygen.

An unprotected metal surface can be caused by three conditions:

- The metal surface is bare — for example, following an acid cleaning.
- The metal surface is covered with a marginally protective, or nonprotective, iron oxide, such as hematite, Fe_2O_3 (red).
- The metal surface is covered with a protective iron oxide, such as magnetite, Fe_3O_4 (black), but holidays or cracks exist in the coating.

Breakdown, or cracking of the magnetite, is due largely to mechanical and thermal stresses induced during normal boiler operation. These stresses are increased — and, therefore, are more damaging — during boiler start-up, during boiler shutdown, and during rapid load swings. During normal boiler operation, the environment favors rapid repair of breaches in the magnetite. However, if excessive levels of oxygen are present, either during operation or outages, the cracks in the magnetite cannot be adequately repaired and corrosion commences.

Identification

Simple visual examination is sufficient if affected surfaces are accessible. Nondestructive testing techniques, such as ultrasonic testing, may be required if affected surfaces are not accessible.

Elimination

The three critical factors that govern oxygen corrosion in a boiler are moisture or water, oxygen, and an inadequately protected metal surface.

Operating boiler

Water is always present in an operating boiler. Also, the protective magnetite coating exists in a state of continuous breakdown and repair. At any given time, holidays and cracks in the magnetite will be present, although the percentage of the entire internal surface they represent will be very small. Therefore, since both water and corrosion sites are present, mitigation of oxygen corrosion is achieved by sufficiently diminishing dissolved oxygen levels.

Possible causes of excessive levels of dissolved oxygen are, for example, a malfunctioning deaerator, improper feed of oxygen-scavenging chemicals, or air in-leakage. (See the section titled Sources of Air In-Leakage in this chapter.) Monitoring of oxygen levels at the economizer inlet, especially during start-up and low-load operation, is recommended.

Idle boiler — wet lay-up

An idle boiler during wet lay-up is subject to conditions similar to those in an operating boiler as far as oxygen corrosion is concerned. Therefore, the preventive method, reduction of oxygen content to very low levels, and continuous control that prevents these levels from rising, is the same. In general, this procedure requires complete filling of the boiler, use of sufficiently high levels of oxygen-scavenging chemicals, and maintenance of properly adjusted pH levels, as well as periodic water circulation.

Idle boiler — dry lay-up

Successful protection of an idle boiler during dry lay-up depends upon consistent elimination of moisture and/or oxygen. A procedure for boiler protection by dry lay-up can involve the use of desiccants and nitrogen blankets, or the continuous circulation of dry, dehumidified air (<30% relative humidity).

Boiler after chemical cleaning

Protection of a boiler following acid cleaning is achieved by developing a protective iron oxide coating on the metal surface. This is usually accomplished by a thorough rinsing followed with a "post boilout." A sodium carbonate solution or other alkaline substance can be used in the post-boilout–passivation step.

Cautions

The knoblike mound of corrosion products that frequently covers pit sites is sometimes misidentified as simple deposits. Correct identification of these mounds, known as *tubercles,* can be achieved through a consideration of the following:

- Tubercles will overlie corrosion sites. Under the influence of sufficient fluid velocity the tubercle may be elongated in the flow direction.

- A tubercle is highly structured and usually consists of a hard, brittle, outer shell of reddish corrosion products that encapsulates an inner core of soft, voluminous, dark corrosion products.

Sources of Air In-Leakage

Low-pressure feedwater heaters are often under negative pressure during low-load operation, allowing air to be drawn into the feedwater through leaky valves, pumps, flanges, etc. Other sources of air in-leakage are shown in Table 8.1.

TABLE 8.1   Sources of Air In-Leakage

Location	Leakage source
Turbine	Gland seals
	Flanges, hood, and expansion joints
	Penetrations
	Valves
Condenser	Penetrations and flanges
	Boot
	Manways
	Hot-well sight glass
Condensate pump suction	Valves, gauges, flanges
	Pump-shaft seal
Feedwater heater	Valves
Other	Perforated explosion diaphragms
	Valves
	Expansion joints
	Flash tanks

Related Problems

See also Chap. 15, "Corrosion-Fatigue Cracking."

CASE HISTORY 8.1

Industry:	Pulp and paper
Specimen Location:	Superheater
Specimen Orientation:	Horizontal, vertical
Years in Service:	10
Treatment Program:	Dry lay-up using nitrogen blanket
Tube Specifications:	2 in. (5.1 cm) outer diameter

The oxygen pits illustrated in Figs. 8.1 and 8.4 were discovered during a hydrotest of a recovery boiler that had been in wet lay-up over the summer. The superheater section had been in dry lay-up under a nitrogen blanket.

Gravity-induced drainage of corrosion products from sites on vertical sections is apparent in Fig. 8.1. Elliptical rings surrounding the pit on the horizontal portion of the pendant section indicate that a pool of condensate had been present (Fig. 8.4). Close examination under a low-power stereoscopic microscope revealed contraction cracks in the protective magnetite at the pit sites. It is apparent that, despite the precautions of dry lay-up, the tube had been exposed to condensed moisture and atmospheric oxygen.

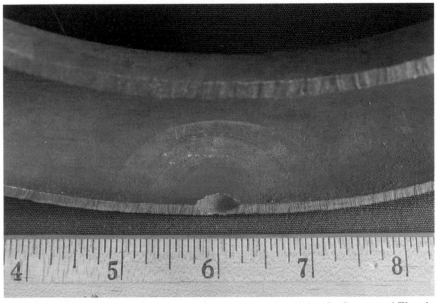

Figure 8.4 Elliptical water rings surrounding oxygen pit in a U-bend. *(Courtesy of Electric Power Research Institute.)*

CASE HISTORY 8.2

Industry:	Pulp and paper
Specimen Location:	Convection bank
Specimen Orientation:	Various
Years in Service:	15
Treatment Program:	Wet lay-up, hydrazine, morpholine
Drum Pressure:	700 psi (4.8 MPa)
Tube Specifications:	2 in. (6.3 cm) outer diameter

Figures 8.5 through 8.7 illustrate typical oxygen pitting resulting from an improperly executed wet lay-up. Inadequate feed of oxygen-scavenging chemicals, and/or failure to maintain proper water circulation, may have been responsible.

Gravitationally induced drainage of corrosion products is apparent in Fig. 8.5. This is evidence of stagnant or very low flow of water during lay-up. Significant water flow would have reoriented the drainage lines in the direction of water flow. Typical knoblike mounds of iron oxide corrosion products (tubercles) are shown in Figs. 8.6 and 8.7.

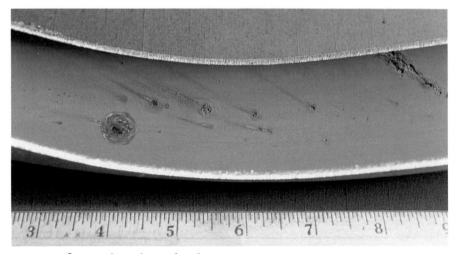

Figure 8.5 Oxygen pits on internal surface.

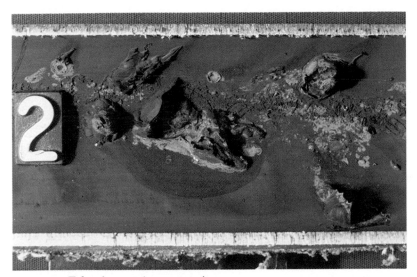

Figure 8.6   Tubercles covering oxygen pits.

Figure 8.7   Tubercles covering oxygen pits.

CASE HISTORY 8.3

Industry:	Steam heating
Specimen Location:	Fire-tube boiler
Specimen Orientation:	Horizontal
Years in Service:	4
Water-Treatment Program:	Polymer and sodium sulfite
Tube Specifications:	3 in. (7.6 cm) outer diameter

Figure 8.8 illustrates the external surface of a fire-tube boiler that has sustained severe oxygen corrosion. Interruptions of oxygen-scavenger feed lasting as long as 1 week had occurred. It is probable that the pitting initiated and progressed during these interruptions. Deposits, which can also induce pitting (deposit corrosion), were not present on the tube.

Figure 8.8 Oxygen pitting on the external surface of fire tube.

CASE HISTORY 8.4

Industry:	Steam heating
Specimen Location:	Condensate return line
Specimen Orientation:	Horizontal
Years in Service:	9
Water-Treatment Program:	Neutralizing amine
Tube Specifications:	1¼-in. (3.2 cm) outer diameter

Figure 8.9 illustrates a recurring problem in one region of the condensate system of a boiler that operated 8 to 10 hours per day. The reddish color of the iron oxides and the tubercles covering the pit sites identify the deterioration as oxygen pitting.

The condensate system is pressurized during service and carries water at 220°F (104°C). Although oxygen could be present in the system under these conditions, it is more probable that the source of oxygen is air sucked into the line as the system cooled during idle periods. A treatment utilizing filming amine, rather than a simple neutralizing amine would probably be more successful in mitigating corrosion of this type.

Figure 8.9 Internal surface of condensate line.

CASE HISTORY 8.5

Industry:	Chemical process
Specimen Location:	Economizer
Specimen Orientation:	Horizontal
Years in Service:	7
Water-Treatment Program:	Polymer and oxygen scavenger
Drum Pressure:	600 psi (4.1 MPa)
Tube Specifications:	2½ in. (6.3 cm) outer diameter

The reddish color of the iron oxides and the presence of tubercles capping iron oxide–filled pits (Fig. 8.2) is typical of exposure of economizer steel to water containing excessively high levels of dissolved oxygen. Pitting and perforation of economizer tubes was a recurrent problem at this plant. Failures were occurring every 3 or 4 months.

Oxygen content of the water was measured at 5 to 9 ppb. Excursions to higher levels were suspected but could not be documented. The boiler was operated continuously. Although the source of the oxygen was not identified, it is clear that excessively high levels existed in the affected regions of the economizer.

Fire-Side Corrosion

Introduction

In simple terms, combustion involves the rapid reaction of oxygen with the basic chemical elements in fuels — carbon, hydrogen, and sulfur — with a consequent release of heat and the formation of combustion products (Fig. 9.1). "Foreign" material present in fuel forms the combustion by-product referred to as ash.

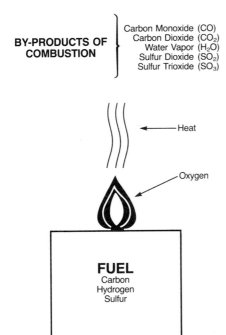

BY-PRODUCTS OF COMBUSTION
Carbon Monoxide (CO)
Carbon Dioxide (CO_2)
Water Vapor (H_2O)
Sulfur Dioxide (SO_2)
Sulfur Trioxide (SO_3)

Heat

Oxygen

FUEL
Carbon
Hydrogen
Sulfur

Figure 9.1 Combustion products resulting from burning of fuels.

Regardless of the original physical state of the fuel, combustion may convert fuel components to any or all of the three states of matter — solid, liquid, or gas. In a combustion device, the flue-gas temperature may range from 3000°F (1650°C) in the flame to 300°F (121°C) or less at the exhaust stack. As combustion products cool on their way to the exhaust stack, gaseous products may condense to liquids, and liquids may freeze to solids. This cooling may occur rapidly on heat-transfer surfaces that are cold relative to the flue gas, such as boiler and superheater tube walls.

In addition, combustion products rarely remain as individual oxides, but generally interact to form new families of compounds and complexes. At times, these new substances may have fusion temperatures that are lower than those of the substances from which they were formed. The presence of these liquid substances may be responsible for difficulties with fire-side corrosion.

Chapters 9 through 13 of this book discuss various types of fire-side corrosion.

Oil-Ash Corrosion

Locations

Oil-ash corrosion is a high-temperature, liquid-phase corrosion phenomenon generally occurring where metal temperatures are in the range of 1100 to 1500°F (593 to 816°C). It is found in superheater and reheater sections of the boiler, especially utility boilers. It may affect the tubes, which are cooled, or it may affect support and attachment equipment, which operates at higher surface temperatures than the tubes.

General Description

Fire-side corrosion may become a problem when fuel supply or fuel type is changed. This change may result in the formation of an "aggressive" ash. Oil-ash corrosion occurs when molten slag containing vanadium compounds forms on the tube wall according to the following sequence:

1. Vanadium compounds and sodium compounds present in the fuel are oxidized in the flame to V_2O_5 and Na_2O.
2. Ash particles stick to metal surfaces, with Na_2O acting as a binding agent.
3. $V_2O_5 + Na_2O$ react on the metal surface, forming a liquid (eutectic).
4. The liquid formed fluxes the magnetite, exposing the underlying metal to rapid oxidation.

It is believed that corrosion occurs by catalytic oxidation of the metal by vanadium pentoxide (V_2O_5) or complex vanadates. The resulting rapid oxidation of the metal then reduces wall thickness, which, in turn, reduces load-carrying area. This reduction in load-carrying area results in an increase in stresses through the thinned region. The combined influence of increased stress level and high metal temperatures eventually results in failure by creep rupture.

Corrosion of superheater and reheater tubes due to slag having fusion temperatures in the 1100 to 1300°F (593 to 704°C) range was largely responsible for the deviation from the trend toward higher steam temperatures that occurred in the early 1960s. Practically all utility-boiler installations are now designed for maximum steam temperatures in the 1000 to 1025°F (538 to 551°C) range.

Critical Factors

A corrosive slag may develop when fuel oil that contains high levels of vanadium, sodium, or sulfur, or a combination of these elements, is used; when excessive amounts of excess air are available for the formation of V_2O_5; or when metal temperatures exceeding 1100°F (593°C) are achieved. As the temperature increases, the range of compositions of $Na_2O \cdot V_2O_5$ that forms liquids expands considerably (Fig. 9.2).

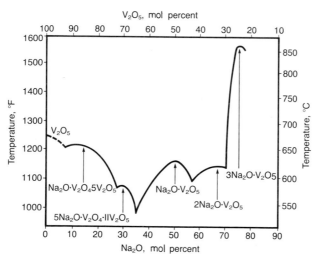

Figure 9.2 Relationship between temperature and compositions of liquid forms involving Na_2O and V_2O_5. *(Courtesy of William T. Reid,* External Corrosion and Deposits: Boilers and Gas Turbines, *American Elsevier, New York, 1971, p. 137.)*

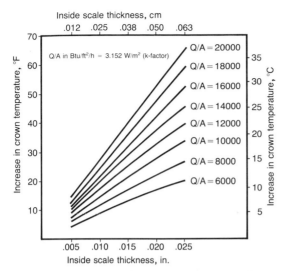

Figure 9.3 Relationship between increase in metal temperature and thickness of internal scale for steam-cooled tubes. *(Courtesy of John Wiley and Sons,* Metallurgical Failures in Fossil Fired Boilers, *by David N. French, New York, 1983.)*

Figure 9.3 indicates that as the thickness of internal scale increases, the metal temperature also increases, since the scale acts as thermal insulation. Hence, in older units, which may have established relatively thick layers of internal scale, the metal temperature will increase and may exceed temperatures at which sodium–vanadium complexes form liquids. If this occurs, sudden, unexpected problems with oil-ash corrosion may appear, even though operating parameters and fuel chemistry remain unchanged.

Identification

Figure 9.4 illustrates the appearance of oil-ash corrosion on a low-alloy steel tube. A section of a stainless steel reheater tube that has suffered oil-ash corrosion is illustrated in Fig. 9.5. Figure 9.6 illustrates the deterioration of a stainless steel tube at an attachment. The attachment, which protruded into the gas stream, acted as a heat-transfer fin, causing metal temperatures at its base to increase. Severe oil-ash corrosion resulted wherever the metal temperature exceeded 1100°F (593°C).

Figure 9.4 Wall thinning from oil-ash corrosion. Corrosion rates of 30 mil/y (0.76 mm/y) have been observed. *(Courtesy of Electric Power Research Institute.)*

Elimination

The first step in combating oil-ash corrosion is chemical analysis of both the fuel and ash to determine whether corrosive constituents are present. It is also important to know the fusion temperature of the ash. Annual surveys of tube-wall thickness using ultrasonic testing can give early warning of impending problems. If tube failure occurs, a wall-thickness survey can determine the extent and severity of the problem.

Elimination of oil-ash corrosion is accomplished by controlling the critical factors that govern it. First, if fuels containing very low quantities of vanadium, sodium, and sulfur cannot be specified, then recommendation of a fuel-treatment additive to prevent the formation of low-melting eutectics may be necessary. The use of magnesium compounds has proven to be

Figure 9.5 Oil-ash corrosion on stainless steel reheater tube. *(Courtesy of John Wiley and Sons,* Metallurgical Failures in Fossil Fired Boilers, *by David N. French, New York, 1983.)*

Figure 9.6 High-temperature corrosion at the base of an attachment (top of photograph) on a stainless steel reheater tube. *(Courtesy of Electric Power Research Institute.)*

economically successful in mitigating problems of oil-ash corrosion. Magnesium forms a complex with vanadium ($3MgO \cdot V_2O_5$) whose fusion temperature is significantly above that attained in most superheater and reheater sections. Second, the boiler should be fired with low excess air to retard V_2O_5 formation. Third, the superheater and reheater metals should be prevented from exceeding 1100°F (593°C). Boilers having drainable superheater and reheater sections should be chemically cleaned periodically to prevent excessive buildup of internal scale.

Coal-Ash Corrosion

Locations

Coal-ash corrosion is a high-temperature, liquid-phase corrosion phenomenon that occurs on metals whose surface temperatures are in the range of 1050 to 1350°F (566 to 732°C). It is usually confined to superheater and reheater sections of the boiler.

General Description

Coal-ash corrosion may occur when the fuel supply or fuel type is changed, resulting in production of an aggressive ash.

During coal combustion, minerals in the coal are exposed to high temperatures, causing release of volatile alkali compounds and sulfur oxides. Coal-ash corrosion occurs when fly ash deposits on metal surfaces where temperatures range from 1050 to 1350°F (566 to 732°C). With time, the volatile alkali compounds and sulfur compounds condense on the fly ash and react with it to form complex alkali sulfates such as $K_3Fe(SO_4)_3$ and $Na_3Fe(SO_4)_3$ at the metal/deposit interface. The molten slag fluxes the protective iron oxide covering of the tube, exposing the metal beneath to accelerated oxidation.

Reduction of wall thickness by this corrosion mechanism effectively increases stresses through the thinned wall. These increased stresses, coupled with high metal temperatures, may lead to final failure by stress rupture.

Critical Factors

The critical factors causing coal-ash corrosion include the use of a coal that produces an aggressive ash, and conditions that produce metal temperatures in the range of 1050 to 1350°F (566 to 732°C).

Identification

Coal-ash corrosion is identified by slag buildup on the tube wall and the associated metal loss. Austenitic stainless steel tubes may exhibit a pockmarked surface appearance. Low-alloy carbon steel tubes typically show a pair of flat zones of metal loss that are located on both sides of the leading face of the tube at orientations of 30 to 45°. Corroded surfaces have a grooved or roughened appearance (Fig. 10.1).

Usually, if corrosion occurs, it will be greatest in tubes having the highest steam temperatures. The highest corrosion rates are generally found on the outlet tubes of radiant superheater or reheater platens.

The corrosion rate is a nonlinear function of temperature, reaching a maximum between 1250 and 1350°F (677 and 732°C). At higher temperatures, the corrosion rate decreases rapidly because of thermal decomposition of the corrosives (Fig. 10.2).

Corrosion is almost always associated with a sintered or slag-type deposit that is strongly bonded to the metal surface. This deposit consists of three distinct layers (Fig. 10.3). The outer layer is a bulky layer of porous

Figure 10.1 Superheater tube (ASME SA-213, grade T22) that ruptured due to thinning by coal-ash corrosion. *(Reprinted with permission of American Society for Metals Handbook, vol. 10, 8th ed., Metals Park, Ohio, 1975, p. 537.)*

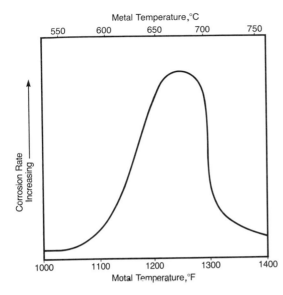

Figure 10.2 Nonlinear relationship between temperature and corrosion rate.

fly ash; the intermediate layer consists of whitish, water-soluble alkali sulfates, which are responsible for the corrosive attack. This layer is typically ⅟₃₂ to ¼ in. (0.79 to 6.35 mm) thick; a thin inner layer is composed of glassy black iron oxides and sulfides at the metal surface. This layer is seldom thicker than ⅛ in. (3.2 mm).

Ultrasonic thickness measurements taken at angles of 30 to 45° on both sides of the leading face of the tube should indicate whether significant metal loss has occurred.

Coal-ash corrosion can occur with any bituminous coal but is more probable with coals containing more than 3½% sulfur and ¼% chlorine.

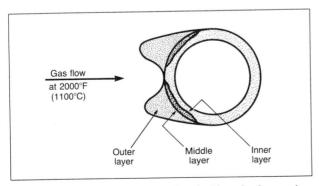

Figure 10.3 Layers of deposit associated with coal-ash corrosion.

Elimination

Generally, a chemical analysis of fuel and ash deposits is recommended to determine what corrosive constituents may be present. A determination of ash-fusion temperatures may also prove valuable.

If corrosion is known to have occurred, ultrasonic thickness surveys are useful for determining the extent and severity of the attack. If corrosion is not severe, economic solutions may include periodic tube replacement, specification of thicker tube walls, use of thermal spray coatings, or pad welding. The use of stainless steel shields over affected areas has extended tube life.

Where coal-ash corrosion is severe, cladding of tubes with a resistant alloy may be required. To date, fuel additives have not been economically successful in combating coal-ash corrosion.

In view of the critical factors that govern this form of corrosion, it may be valuable to blend coals to reduce the percentage of corrosive constituents, and to lower metal temperatures. Methods of lowering metal temperature include lowering steam temperatures, periodic cleaning of drainable superheater and reheater sections to prevent buildup of internal scale, and redesign of affected areas to reduce heat-transfer rates.

CASE HISTORY 10.1

Industry:	Utility
Specimen Location:	Secondary-superheater outlet
Specimen Orientation:	Vertical
Years in Service:	21
Water-Treatment Program:	All volatile treatment
Drum Pressure:	2500 psi (17.2 MPa)
Steam Temperature:	1050°F (566°C)
Tube Specifications:	2⅛ in. (5.4 cm) outer diameter, austenitic (SS321) stainless steel
Fuel:	Eastern coal, 10% ash, 7% S

The boiler from which the tubes illustrated in Figs. 10.4 and 10.5 were removed had experienced chronic corrosion problems in the outlet pendants of the secondary superheater. Ten separate failures had occurred over a 3-year period. The boiler was operated continuously.

The tube submitted for analysis had not failed, but had sustained a 1-in.-wide (2.5-cm-wide) band of severe corrosion along one side of its external surface. Measurement indicated a reduction in tube-wall thickness of 0.085 in. (2.2 mm).

Figure 10.4 Contour of external surface in corroded zone. Note pockmarked sur face appearance characteristic of attack on stainless steel.

Figure 10.5 Transverse profile of tube showing metal loss.

Microstructural examinations revealed solidified eutectics of slag covering the external surface in the corroded zone. Chemical analysis of material covering the corroded region revealed large amounts of iron, sulfur, and potassium, as well as components of fly ash.

Evidence from visual, microstructural, and chemical analyses identified the wastage as coal-ash corrosion resulting from the formation of complex alkali sulfates at the metal surface.

Waterwall–Fire-Side Corrosion

Locations

As the title of this chapter indicates, the type of corrosion next discussed affects the fire side of waterwalls in coal-fired boilers. Boilers with low burner-to-side-wall or burner-to-rear-wall clearances are often affected. Waterwall–fire-side corrosion is frequently found in the wind-box area or burner area.

General Description

Waterwall–fire-side corrosion may develop when incomplete fuel combustion occurs (reducing conditions). Incomplete combustion causes release of volatile sulfur compounds, which may form pyrosulfates. Sodium and potassium pyrosulfates ($Na_2S_2O_7$ and $K_2S_2O_7$) may have melting points of 800°F (427°C) or less. Both have a high chemical activity. These molten slags may flux the protective magnetite on tube surfaces, causing accelerated metal deterioration along the crown of the tube.

Depending on the amount of SO_3 in the salt, the melting point may be less than 770°F (396°C) for $K_2S_2O_7$ and around 754°F (387°C) for $Na_2S_2O_7$ (Fig. 11.1). As temperature increases, the amount of SO_3 required to form a liquid phase also increases significantly. Consequently, neither sodium nor potassium pyrosulfates are likely to be present as liquid except on relatively cool surfaces, such as the waterwalls.

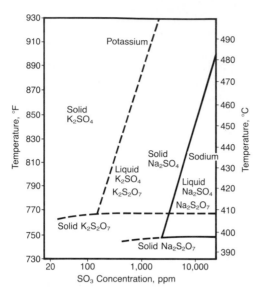

Figure 11.1 Relationships between temperature and SO_3 concentration to produce solid and liquid phase in the $Na_2SO_4 \cdot SO_3$ and $K_2S_4 \cdot SO_3$ systems. (Reprinted with permission from William T. Reid, External Corrosion and Deposits: Boilers and Gas Turbines, *American Elsevier, New York, 1971,* p. 106.)

Critical Factors

Insufficient oxygen in the burner zone is a primary factor in development of waterwall corrosion. Poor combustion conditions and steady or intermittent flame contact with the furnace walls, combined with coals that are capable of forming an ash with a low fusion temperature, produce a hot, fuel-rich corrosion environment.

Identification

Waterwall–fire-side corrosion is frequently characterized by metal loss along the crown of the tube and may extend uniformly across several tubes in a particular location (Fig. 11.2). Corroded regions may be covered by abnormally thick layers of iron oxide and iron sulfide corrosion products. At times, flow patterns of the liquid pyrosulfates are apparent in regions of metal wastage.

Chemical analysis of both fuel and ash may be required to determine the level of corrosive substances. A determination of the completeness of combustion can also indicate whether corrosion is occurring. Irregular combustion, malfunctioning burners, long flames, and high carbon content of the ash (3% or more), indicate that waterwall corrosion may be occurring.

In suspect areas, the use of ultrasonic thickness surveys may determine whether corrosion is occurring as well as the extent and rate of corrosion.

Figure 11.2 Severe fire-side metal loss on waterwall tubes. *(Courtesy of Electric Power Research Institute.)*

Elimination

Since the critical factors governing this form of deterioration relate to combustion characteristics, reduction or elimination of this problem can be realized by making appropriate changes in the combustion process. Changes might consist of improving burning efficiency by grinding coal to a finer, more uniform size; balancing fuel supply to individual burners; adjusting burners to prevent flame impingement; and increasing and redistributing secondary air.

Experience has shown, however, that only marginal improvement in combustion can be expected from these corrections. A furnace modification may be necessary to achieve substantial improvement. To date, fuel additives have not been economically successful in combating waterwall–fire-side corrosion.

Ultrasonic thickness surveys are used to determine extent and severity of existing damage. If damage is minor, patch welding of affected areas may be satisfactory. If damage is severe, installation of thicker tubes or of corrosion-resistant alloy tubes, use of thermal spray coatings, cladding of tubes, and the use of shields may be economically justifiable.

CASE HISTORY 11.1

Industry:	Utility
Specimen Location:	Roof tube
Specimen Orientation:	Horizontal
Years in Service:	35
Water-Treatment Program:	Congruent control
Drum Pressure:	2150 psi (14.8 MPa)
Tube Specifications:	3 in. (7.6 cm) outer diameter, studded
Fuel:	Coal, 13% ash

Six tube failures of the type illustrated in Fig. 11.3 occurred over a period of several weeks. Only roof tubes had failed.

Visual examination of the external surface of one of the tubes revealed a deep, longitudinal fissure at the base of a stud (Fig. 11.3). This surface was covered with a hard, tenacious, light-colored slag. Chemical analysis of this fire-side slag revealed 42% sulfur and 18% sodium. The pH of a 1% slurry of the deposit was 2.9.

Microstructural examinations revealed no thermal alteration of the tube metal. However, deep, intergranular fissures filled with a complex sulfate eutectic were observed originating on the external surface.

Figure 11.3 Longitudinal fissure adjacent to stud; fissure was caused by the combined effects of stresses imposed by internal pressure and the presence of a molten phase.

Visual and microstructural examinations, coupled with chemical analysis of the slag, revealed that the cracking apparent in Fig. 11.3 resulted from penetration of molten sodium pyrosulfate along grain boundary pathways of the tube-wall metal during boiler operation. Stresses imposed by normal internal pressure acted synergistically with the molten slag to produce the intergranular penetration. Microstructural examinations reveal that this attack was localized to a small region around the primary crack.

The presence of sodium pyrosulfates indicates that reducing conditions or incomplete combustion existed in the firebox, possibly from insufficient oxygen in the burner zone or unsatisfactory grinding of the coal. Insufficient oxygen in the burner zone may be caused by insufficient excess air.

CASE HISTORY 11.2

Industry:	Pulp and paper
Specimen Location:	Adjacent to primary air port, recovery boiler
Specimen Orientation:	Vertical
Years in Service:	10
Tube Specifications:	3 in. (7.6 cm) outer diameter, stainless steel clad, carbon steel tube
Fuel:	Black liquor

Figure 11.4 shows metal loss from the external surface near a longitudinally oriented fin. Metal loss was confined to an elliptical region centered on the fin. The stainless steel cladding as well as some underlying carbon steel was corroded away.

The corroded region was covered with a layer of brown corrosion product and deposits. Chemical analysis of this layer revealed a content of 52% iron, 24% sodium, and 14% carbonate.

Visual and microstructural evidence, coupled with analysis of the corrosion products, indicated that metal loss was caused by exposure of the metal to a molten salt of sodium. The fusion temperature of this salt may have been depressed by the presence of the carbonate.

Rates of metal loss from corrosion of this type vary depending on metal temperature and furnace design. Corrosion rates of 30 mil/y (0.76 mm/y) have been reported.

Mitigation of this problem requires redesign of tube openings so that seals are tight to flue gas. In addition, crevices where corrosive substances can concentrate must be eliminated.

Repair of damage is most successful when weld overlays of high-nickel stainless steels and thermal spray coatings are used.

Figure 11.4 Metal loss around a fin resulting from exposure of a stainless steel–clad tube to a molten sodium salt.

CASE HISTORY **11.3**

Industry:	Pulp and paper
Specimen Location:	Wall tube, recovery boiler
Specimen Orientation:	Vertical
Years in Service:	13
Water-Treatment Program:	Polymer
Drum Pressure:	875 psi (6 MPa)
Tube Specifications:	2½ in. (6.3 cm) outer diameter
Fuel:	Black liquor

The corrosion apparent in Figs. 11.5 and 11.6 occurred over a very small area of the wall at a position 40 ft (12.2 m) above the floor and 4 ft (1.2 m) above oil guns. A failure of this type had not occurred previously.

This sample had been water-washed before removal, which reportedly may have removed a layer of frozen smelt. The fire side of this boiler is water-washed twice per year.

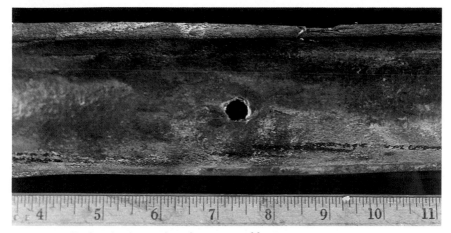

Figure 11.5   Perforation in a region of severe metal loss.

Figure 11.6   Severe metal loss along crown of tube resulting from exposure to a molten phase.

The ⅜-in. hole in Fig. 11.5 is centered in a region of severe external corrosion along the crown of the tube. The corrosion has produced deep, broad depressions, which give the surface a rolling contour. Microstructural examinations revealed no evidence of overheating.

The fact that this corrosion was localized in the region of the oil guns suggests that a reducing environment, created locally during the operation of the guns, produced a corrosive molten phase having a low melting temperature. In this case, corrosion would occur only during oil-gun use, which is intermittent. However, the accumulated corrosion over 13 years of service was apparently sufficient to result in failure.

Chapter 12

Cold-End Corrosion during Service

Locations

Cold-end corrosion will occur wherever the temperature of metal drops below the sulfuric acid dew point of the flue gas. Most problems caused by cold-end corrosion occur in relatively low temperature boiler components such as the economizer, air preheater, induced-draft fan, flue-gas scrubbers, and stack.

General Description

In most combustion systems, flue-gas temperatures can range from 3000°F (1650°C) in the flame to 250°F (121°C) or less at the stack. This temperature change can cause numerous chemical and physical changes in the components of the flue gas. Among the most troublesome changes is the reaction between water vapor and sulfur trioxide to form sulfuric acid.

As flue gas cools, vapor-phase sulfuric acid forms. If the sulfuric acid vapor contacts a relatively cool surface, it may condense as liquid sulfuric acid. The temperature at which sulfuric acid first condenses (sulfuric acid dew point) varies from 240 to 330°F (116 to 166°C) or higher, depending on sulfur trioxide and water-vapor concentrations in the flue gas. The corrosion resulting from condensation of sulfuric acid on metal surfaces is termed "cold-end corrosion" because it generally affects the cooler regions of the combustion system.

In general, the problem is associated with the combustion of fuel that contains sulfur or sulfur compounds. Sulfur in the fuel is oxidized to sulfur trioxide in the following sequence:

$$S + O_2 \rightarrow SO_2$$

A small fraction (1% to 3%) of the sulfur dioxide produced is additionally oxidized to sulfur trioxide by direct reaction with atomic oxygen in the flame:

$$SO_2 + O \rightarrow SO_3$$

Catalytic oxidation to SO_3 is also possible if ferric oxide, vanadium pentoxide, or nickel is present:

$$SO_2 + \tfrac{1}{2}O_2 + \text{catalyst} \rightarrow SO_3$$

The quantity of sulfur trioxide and moisture in the flue gas affects the temperature at which the dew point is reached. The graph in Fig. 12.1 illustrates the general relationship between sulfur trioxide concentration and dew point.

Corrosion may occur wherever metal temperatures are less than the

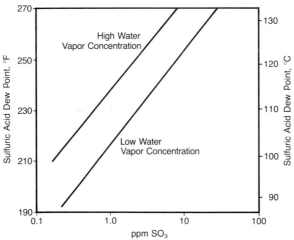

Figure 12.1   Relationship between SO_3 levels in flue gas and sulfuric acid dew point.

sulfuric acid dew point. Below this dew point, sulfuric acid forms on metal surfaces and corrodes the metal according to the following reaction:

$$H_2SO_4 + Fe \rightarrow FeSO_4 + H_2\uparrow$$

Note that it is the temperature of the metal, not the temperature of the flue gas, that is critical. Even though the flue-gas temperature is well above the dew point, corrosion is a distinct possibility wherever metal temperatures are less than the dew point.

Critical Factors

The critical factors governing cold-end corrosion include the presence of corrosive quantities of sulfur trioxide, the presence of moisture in the flue gas, and the presence of metals whose surface temperature is below the sulfuric acid dew point.

The amount of sulfur trioxide produced increases with increases in the level of excess air, gas residence time, gas temperature, amount of catalyst present, and sulfur level in the fuel (Fig. 12.2).

The amount of moisture produced is dependent on many factors. Sources include moisture content in the fuel, fuel combustion, leaks in boiler tubes, and steam from soot blowing.

A metal whose surface temperature is below the sulfuric acid dew point is susceptible to cold-end corrosion. The dew point increases as the quantity of sulfur trioxide in the flue gas and the moisture content of the flue gas increase.

The severity of cold-end corrosion is most extreme with "sour" gaseous fuels, while it is less extreme with oil. Severity of corrosion is least with coal. The combined burning of sulfur- and vanadium-containing fuel oil

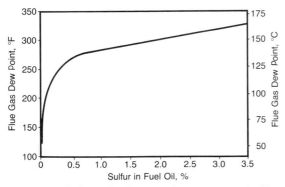

Figure 12.2 Relationship between the percentage of sulfur in fuel oil and flue-gas dew point.

with natural gas is worse than use of oil alone because of the high water-vapor content that results from burning of natural gas.

Identification

Cold-end corrosion frequently produces general, smooth, featureless metal loss. Rough, rust-colored surfaces also may be observed.

Elimination

Elimination of cold-end corrosion is achieved by gaining control of the critical factors that govern it. The critical factors include the presence of corrosive quantities of sulfur trioxide in flue gas, the presence of excessive quantities of moisture in the flue gas, and the presence of metals whose surface temperature is below the sulfuric acid dew point.

To eliminate the presence of corrosive quantities of sulfur trioxide in flue gas it is necessary to operate the boiler at or below 5% excess air to the burners, to minimize air infiltration, and to specify fuels with low sulfur content.

To eliminate the presence of excessive quantities of moisture in the flue gas, it is necessary to specify fuel with low moisture content, to prevent tube leaks, and to reduce the amount of soot blowing.

Substantial design changes are often required to eliminate metals with surface temperatures below the sulfuric acid dew point. These design changes are beyond the scope of this manual.

In boilers equipped with multiple air preheaters it is possible to isolate and perform an on-line washing of one of the preheaters.

Feeding of fuel additives and/or cold-end additives constitutes a chemical solution to this problem.

Cautions

Metal surfaces that have suffered cold-end corrosion may be covered with deposits and/or corrosion products, making visual identification difficult. In addition, cold-end corrosion typically produces smooth, uniform, featureless metal loss, such that the attacked surfaces closely resemble the original, unaffected surface contour. Thickness measurements using ultrasonic techniques serve as a nondestructive means of evaluation. Visual comparison of surface profiles and metal thicknesses on a section cut from the equipment may also reveal the occurrence of cold-end corrosion.

CASE HISTORY 12.1

Industry:	Pulp and paper
Specimen Location:	Economizer, recovery boiler
Specimen Orientation:	Vertical
Years in Service:	20
Drum Pressure:	600 psi (4.1 MPa)
Tube Specifications:	2½-in. (6.3 cm) outer diameter
Fuel:	Black liquor

Figure 12.3 is a close-up photograph of the external surface of an economizer tube that has sustained general corrosion. Note the pockmarked contour. Corrosion occurred in regions of the economizer where sulfuric acid condensed from the flue gas. Such condensation, if it occurs, generally affects the tubes at the end of the economizer where feedwater enters — i.e., the end where metal temperatures are lower.

Figure 12.3 Pockmarked contour of economizer tube exposed to condensed sulfuric acid. (Magnification: 7.5 X.)

CASE HISTORY 12.2

Industry:	Sugar
Specimen Location:	Economizer
Specimen Orientation:	Horizontal
Tube Specifications:	1⅞ in. (4.8 cm) outer diameter
Fuel:	No. 6 fuel oil (1.9% S)

The corrosion and perforations apparent in Fig. 12.4 represent a chronic problem in this boiler. The external surface, including fin surfaces, exhibited smooth, general metal loss. In many cases, severe corrosion along the attachment line had caused the fins to separate from the tube wall.

The deposits and corrosion products covering the corroded surfaces were analyzed by x-ray diffraction and identified as hydrated iron sulfate. The pH of a 1% slurry of the material was measured at 2.3. The corrosion was caused by condensation of sulfuric acid on the cold tube surfaces during boiler operation.

Figure 12.4   Corrosion and perforations of finned economizer tube resulting from exposure to condensed sulfuric acid. Note detachment of fins from tube wall.

Dew-Point Corrosion during Idle Periods

Locations

Dew-point corrosion during idle periods can occur in the economizer, and also anywhere in the boiler where external surfaces can be covered with acid-forming, sulfurous deposits.

General Description

Dew-point corrosion may result in significant corrosion of external metal surfaces during idle periods. As the boiler cools, the temperature of its external surface may drop below the dew point, allowing moisture to form on tube surfaces. Moisture, in combination with sulfurous deposits, may form a low-pH electrolyte that is capable of generating corrosion rates of 500 mil/y (12.7 mm/y). Figure 13.1 illustrates an ash-covered tube with dew-point corrosion before deposits were removed. Figure 13.2 illustrates the corroded surface of the tube after the deposits were removed.

Critical Factors

Dew-point corrosion is caused by the presence of sulfurous ash deposits on tube surfaces and the reduction of metal temperatures within the boiler to

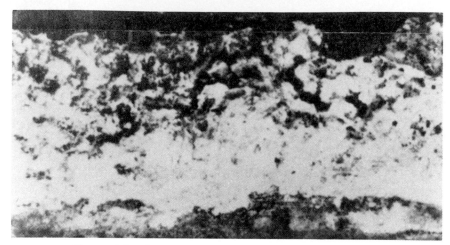

Figure 13.1 Heavy ash buildup on fire side of boiler tube. (*Reprinted with permission from Daniel P. Hamilton, "Preventing Dew-Point Corrosion during Boiler Shutdown,"* Plant Engineering, *July 22, 1982, p. 73.*)

temperatures below the dew point. Sulfurous ash deposits may form a low-pH electrolyte when moisture is present.

Identification

Once deposits are removed, simple visual inspection should disclose the occurrence of dew-point corrosion (Fig. 13.2). Typically, metal attack will

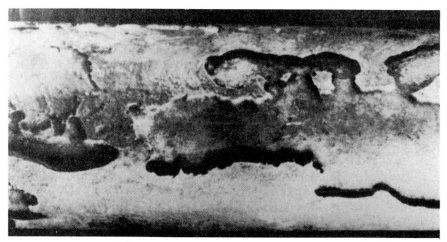

Figure 13.2 Severe, localized corrosion apparent after deposit removal. Areas adjacent to the deposits remain uncorroded. (*Reprinted with permission from Daniel P. Hamilton, "Preventing Dew-Point Corrosion during Boiler Shutdown,"* Plant Engineering, *July 22, 1982, p. 73.*)

be confined to surfaces that were covered with the sulfurous ash deposits. The attack often produces well-defined regions of metal loss and leaves islands of metal relatively intact.

Analysis of ash deposits covering the tube may also be useful in identifying dew-point corrosion. Generally, dew-point corrosion is caused by exposure to sulfurous deposits which form low-pH water solutions.

Elimination

Elimination is achieved through controlling the critical factors that cause the corrosion. Two critical factors govern dew-point corrosion; the presence of sulfurous ash deposits which form low-pH electrolytes when combined with water, and reduction of fire-side metal temperatures below the dew point.

In order to eliminate the presence of sulfurous ash deposits, the following steps may be taken:

- Specify fuels with lower sulfur content. This reduces or eliminates the formation of corrosive ash.
- Remove fire-side deposition from metal surfaces immediately after boiler shutdown by using high-pressure water sprays. This should be followed by a lime wash to neutralize remaining acidic substances. Allow the metal surfaces to air-dry.
- Fire-side metal can be protected from rust by coating metal surfaces with a light oil.
- Containers of unslaked lime placed in the firebox during shutdown will help keep the air dry. This lime should be renewed periodically.

It is generally not possible to maintain metal temperatures above the dew point unless the boiler-water temperature is maintained above the dew point.

Cautions

Corrosion may not be visually apparent until deposits are removed. The irregular pattern of metal loss typical of dew-point corrosion may be helpful in distinguishing it from cold-end corrosion.

Related Problems

See Chap. 12, "Cold-End Corrosion during Service."

CASE HISTORY 13.1

Industry:	Pulp and paper
Specimen Location:	Screen tube, recovery boiler
Specimen Orientation:	Vertical
Years in Service:	10
Drum Pressure:	600 psi (4.1 MPa)
Tube Specifications:	2 in. (5.1 cm) outer diameter
Fuel:	Black liquor

Figure 13.3 illustrates the appearance of the external surface of a screen tube. General metal loss has produced an irregular surface contour. Microstructural examinations revealed that the external surface was covered with a dense layer of iron oxide beneath a layer of iron sulfide. This corrosion occurred during idle periods when moisture, possibly from inadequate water washing, combined with external, sulfur-containing deposits, forming an acidic solution.

Figure 13.3 Appearance of the external surface of a screen tube after exposure to moisture and corrosive deposits during idle periods.

CASE HISTORY 13.2

Industry:	Pulp and paper
Specimen Location:	Front row of economizer, recovery boiler
Specimen Orientation:	Bend, slanted
Years in Service:	11
Drum Pressure:	600 psi (4.1 MPa)
Tube Specifications:	2 in. (5.1 cm) outer diameter
Fuel:	Black liquor

The economizer tube shown in Fig. 13.4 has an irregular, pebblelike surface contour covered with nonprotective corrosion products and deposits. The appearance of this surface is characteristic of corrosion occurring during idle periods when acid-producing salts combine with atmospheric moisture to produce a corrosive environment.

Figure 13.4 External surface of economizer tube following dew-point corrosion.

CASE HISTORY 13.3

Industry:	Pulp and paper
Specimen Location:	Wall tube, 2 ft below roof
Specimen Orientation:	Vertical
Years in Service:	12
Drum Pressure:	600 psi (4.1 MPa)
Tube Specifications:	3 in. (7.6 cm) outer diameter
Fuel:	Black liquor

The corrosion illustrated in Fig. 13.5 was typical of that found in the boiler in the upper furnace area around openings. All corrosion had occurred under refractory. The boiler was water-washed twice annually and was not dried after washing. Measurement revealed a 16% reduction in tube-wall thickness.

The corrosion occurred as a result of the presence of sulfur-containing deposits on the metal surface that were not removed during water washes. Corrosion of this type is not uncommon in stagnant areas, such as beneath refractory. Corrosion rates of up to 50 mil/y (1.27 mm/y) have been reported in similar circumstances.

Figure 13.5   Idle-time corrosion occurring beneath refractory.

CASE HISTORY 13.4

Industry:	Pulp and paper
Specimen Location:	Generating tube 2 ft above mud drum
Specimen Orientation:	Vertical
Years in Service:	21
Drum Pressure:	750 psi (5.2 MPa)
Tube Specifications:	2½ in. (6.3 cm) outer diameter
Fuel:	Black liquor

The metal loss apparent in Fig. 13.6 had affected tubes throughout the generating bank, requiring a major boiler rebuild. The problem was discovered after a major leak resulted in emergency shutdown of the boiler. The corrosion had not been observed previously. The external surfaces of the boiler were water-washed every 4 months.

Measurement revealed a reduction in tube-wall thickness of approximately 20%. The corroded surface is rough, but smooth islands of the original surface remain.

Microstructural examinations revealed no evidence of attack of the external surface by high-temperature slag. The corrosion occurred during idle periods. An inadequate water wash of the boiler can leave corrosive, sulfur-containing, acid-producing salts. The remaining salts, coupled with moisture, are capable of producing the type of corrosion apparent on this tube.

Figure 13.6 Corrosion resulting from inadequate water washing of external surfaces.

CASE HISTORY 13.5

Industry:	Steel
Specimen Location:	Downcomer 12 in. from steam drum
Specimen Orientation:	Slanted 45°
Tube Specifications:	2½ in. (6.3 cm) outer diameter
Fuel:	High-sulfur fuel oil

The tube illustrated in Fig. 13.7 was one of several similar tubes discovered during a fire-side inspection. The tube was positioned at a 45° angle in the boiler. The massive, perforating groove was oriented vertically. It was observed that the groove was directly aligned with water dripping from the economizer section above.

An inspection of the economizer section revealed another series of grooved tubes. The grooves in these tubes were aligned with water dripping from the air preheater section above the economizer.

An inspection of the bottom of the air preheater section, which had recently been water-washed, revealed an accumulation of wet, dripping deposits. Water dripping from these deposits had a pH of 2. Analysis of the deposits by x-ray fluorescence disclosed a high level of sulfur.

Incomplete washing of highly corrosive deposits in the air preheater furnished a source of highly corrosive water, which corroded every tube in its drip paths. Tubes in both the economizer and the boiler proper were affected.

From a corrosion-engineering perspective, this case underscores the importance of recognizing boiler systems as a continuum. Environmental conditions existing in one part of the system can have a direct adverse impact on another part of the system that is physically separate and apparently unconnected.

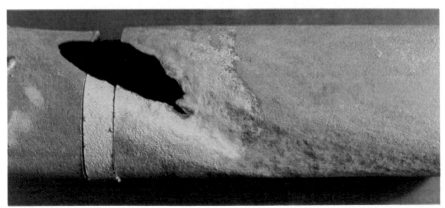

Figure 13.7 Grooving and perforation of a slanted tube resulting from corrosive water dripping from above.

14

Hydrogen Damage

Locations

This form of deterioration is a direct result of electrochemical corrosion reactions in which hydrogen in the atomic form is liberated.* It is typically confined to internal surfaces of water-carrying tubes that are actively corroding.

Generally, hydrogen damage is confined to water-cooled tubes. Damage usually occurs in regions of high heat flux; beneath heavy deposits; in slanted or horizontal tubes; and in heat-transfer regions at or adjacent to backing rings at welds, or near other devices that disrupt flow. Experience has shown that hydrogen damage rarely occurs in boilers operating below 1000 psi (6.9 MPa).

General Description

Hydrogen damage may occur where corrosion reactions result in the production of atomic hydrogen. Damage may result from a high-pH corrosion reaction or from a low-pH corrosion reaction. Damage resulting from a high-pH corrosion reaction is simply caustic corrosion. (Refer to Chap. 4, "Caustic Corrosion.")

---

*"Hydrogen in the atomic form" refers to uncombined atoms of hydrogen (H) as contrasted with molecules of hydrogen (H_2).

Concentrated sodium hydroxide dissolves the magnetic iron oxide according to the following reaction:

$$4NaOH + Fe_3O_4 \rightarrow 2NaFeO_2 + Na_2FeO_2 + 2H_2O$$

With the protective covering destroyed, water is then able to react directly with iron to evolve atomic hydrogen:

$$3Fe + 4H_2O \rightarrow Fe_3O_4 + 8H\uparrow$$

The sodium hydroxide itself may also react with the iron to produce hydrogen:

$$Fe + 2NaOH \rightarrow Na_2FeO_2 + 2H\uparrow$$

If atomic hydrogen is liberated, it is capable of diffusing into the steel. Some of this diffused atomic hydrogen will combine at grain boundaries or inclusions in the metal to produce molecular hydrogen, or will react with iron carbides in the metal to produce methane.

$$Fe_3C + 4H \rightarrow CH_4 + 3Fe$$

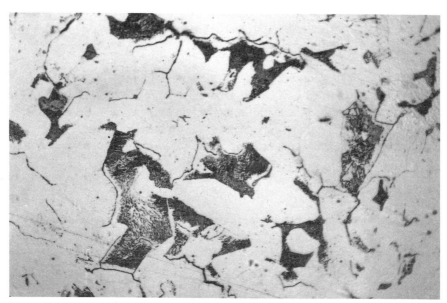

Figure 14.1 Discontinuous intergranular microcracks resulting from methane formation in the grain boundaries. Note decarburization of adjacent pearlite colonies (dark islands). (Magnification: 500×.)

Since neither molecular hydrogen nor methane is capable of diffusing through the steel, these gases accumulate, primarily at grain boundaries. Eventually, gas pressures will cause separation of the metal at its grain boundaries, producing discontinuous intergranular microcracks (Fig. 14.1). As microcracks accumulate, tube strength diminishes until stresses imposed by boiler pressure exceed the tensile strength of the remaining, intact metal. At this point a thick-walled, longitudinal burst may occur (Fig. 14.2). Depending on the extent of hydrogen damage, a large, rectangular section of the wall frequently will be blown out, producing a gaping hole (Fig. 14.3).

Figure 14.2 Thick-walled burst resulting from hydrogen damage. Note areas of gouging adjacent to burst on internal surface. *(Courtesy of Electric Power Research Institute.)*

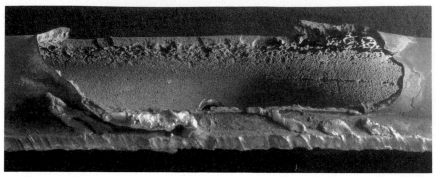

Figure 14.3   Large wall section blown out of a hydrogen-damaged tube from a 2075-psi (14.3-MPa) boiler. Note zone of gouging along fracture line. *(Courtesy of National Association of Corrosion Engineers.)*

Hydrogen damage may also result from a low-pH corrosion reaction in an operating boiler. (Refer to Chap. 6, "Low-pH Corrosion during Service.") Atomic hydrogen may be liberated during corrosion resulting from local low-pH conditions. Atomic hydrogen is capable of diffusing into the metal and reacting to form molecular hydrogen or methane, as described above. Hydrogen damage resulting from exposure to low-pH conditions is mechanistically and physically identical to that resulting from high-pH conditions. The difference is merely the source of the atomic hydrogen.

Critical Factors

The critical factors governing hydrogen damage resulting from high-pH corrosion are identical to those outlined for caustic gouging in Chap. 4, "Caustic Corrosion." The critical factors governing hydrogen damage resulting from low-pH corrosion are identical to those outlined for on-line acid corrosion in Chap. 6, "Low-pH Corrosion during Service."

Identification

It is generally not possible to visually identify hydrogen damage prior to failure. In boilers operating at more than 1000 psi (6.9 MPa), areas that have sustained either high-pH or low-pH corrosion should be considered suspect.

Generally, hydrogen damage is difficult to detect by nondestructive means, although sophisticated ultrasonic techniques have been developed to reveal hydrogen-damaged metal. Ultrasonic thickness checks may disclose corroded areas that should be considered suspect.

Gouging and hydrogen damage resulting from low-pH conditions may be distinguished from damage resulting from high-pH conditions by a consideration of the boiler-water chemistry and the chemistry of the probable sources of contamination. For example, a common source of contamination of boiler water is condenser in-leakage. The source of the cooling water determines whether the in-leakage is acid-producing or base-producing. Fresh water from lakes and rivers usually provides dissolved solids that hydrolyze in the boiler-water environment to form a high-pH substance, such as sodium hydroxide. In contrast, seawater and water from recirculating cooling-water systems incorporating cooling towers may contain dissolved solids that hydrolyze to form acidic solutions.

Elimination

Two critical factors govern susceptibility to hydrogen damage. These are the availability of high- or low-pH substances, and a mechanism of concentration. Both must be present simultaneously for hydrogen damage to occur.

To eliminate the availability of high- or low-pH substances, the following steps should be taken:

Reduce the amount of available free sodium hydroxide.   This can be done in the case of hydrogen damage caused by high pH.

Prevent inadvertent release of regeneration chemicals from makeup-water demineralizers.

Prevent condenser in-leakage.   Because of the powerful concentration mechanisms that may operate in a boiler, in-leakage of only a few parts per million of contaminants may be sufficient to cause localized corrosion and hydrogen damage.

Prevent contamination of steam and condensate by process streams.

Preventing localized concentration of corrosive substances is the most effective means of avoiding hydrogen damage. It is also the most difficult to achieve. Preventing departure from nucleate boiling (DNB), excessive water-side deposition, and the creation of waterlines in tubes may help prevent localized concentration of corrosive substances.

Prevent departure from nucleate boiling.   Preventing DNB usually requires the elimination of hot spots, which is accomplished by controlling the boiler's operating parameters. Hot spots may be caused by excessive overfiring or underfiring, misadjusted burners, change of fuel, gas channeling, and excessive blowdown.

Prevent excessive water-side depositions. To prevent excessive water-side deposition, tube sampling on a periodic basis (usually annually) may be performed to measure relative thickness and amount of deposit buildup on tubes. Tube-sampling practices are outlined in ASTM D887-82. Consult boiler manufacturers' recommendations for acid cleaning.

Prevent waterline formation. Slanted and horizontal tubes are especially susceptible to the formation of waterlines. Boiler operation at excessively low water levels or excessive blowdown rates may create waterlines. Waterlines may also be created by excessive load reduction when pressure remains constant. When load is reduced and pressure remains constant, water velocity in boiler tubes is reduced to a fraction of its full-load value. If it becomes low enough, steam/water stratification occurs and creates stable or metastable waterlines.

Cautions

Hydrogen damage typically produces thick-walled ruptures. Other failure mechanisms producing thick-walled ruptures include stress-corrosion cracking, corrosion fatigue, stress rupture, and, in some rare cases, severe overheating. It may be difficult to visually distinguish ruptures caused by hydrogen damage from other ruptures, although certain features may serve as an aid.

For example, hydrogen damage is almost always associated with metal gouging. (Note cautions listed in Chaps. 4 and 6.) The other failure modes (with the possible exception of corrosion fatigue, which frequently initiates at discrete pits) are not typically associated with gross corrosion.

Tube failures in hydrogen-damaged metal are often manifested as a blowout of a rectangular "window" of the tube wall. This is not a common feature of the other failure modes.

A definitive diagnosis of hydrogen damage may require a formal metallographic examination.

Related Problems

See also Chap. 2, "Long-Term Overheating"; Chap. 3, "Short-Term Overheating"; Chap. 4, "Caustic Corrosion"; Chap. 6, "Low-pH Corrosion during Service"; Chap. 15, "Corrosion-Fatigue Cracking"; and Chap. 16, "Stress-Corrosion Cracking."

CASE HISTORY 14.1

Industry:	Utility
Specimen Location:	Nose section
Specimen Orientation:	Slanted
Years in Service:	25
Water-Treatment Program:	Coordinated phosphate
Drum Pressure:	2075 psi (14.3 MPa)
Tube Specifications:	2 in. (5.1 cm) outer diameter

The boiler from which the section illustrated in Fig. 14.3 was taken is a forced-circulation unit that produces 2½ million pounds (1,134,000 kilograms) of steam per hour. The boiler had been in peaking service for 2 years.

Failures were recurrent in the nose section and roof tubes. An acid cleaning of the boiler was conducted. The failure shown in Fig. 14.3 is one of several that occurred after the acid cleaning.

Note the large, rectangular opening remaining after a section of the tube wall was literally blown out. Fracture edges are thick, blunt, and have an irregular contour. Numerous small, secondary cracks are present along the fracture edges.

Close examination of the illustration reveals internal wall thinning along the fracture surface in the form of intersecting patches of shallow metal loss. In places, the corroded regions are covered with a thick layer of hard, black iron oxide. Where this layer is absent, deep longitudinal cracks can be observed.

Examinations of the microstructure in the fractured region reveal numerous discontinuous intergranular microcracks. Thermal alteration of the microstructure from overheating has not occurred. Iron oxides covering corroded regions of the internal surface are slightly stratified.

The severe hydrogen damage observed in this tube resulted in the blowout of a large section of tube wall. Microstructural evidence (stratification of the iron oxides) suggests that hydrogen damage resulted when the boiler water was contaminated with acid-producing salts (such as chlorides) from condenser in-leakage.

CASE HISTORY 14.2

Industry:	Utility
Specimen Location:	Waterwall
Specimen Orientation:	Vertical
Years in Service:	17
Water-Treatment Program:	Congruent control
Drum Pressure:	2100 psi (14.5 MPa)
Tube Specifications:	2½ in. (6.3 cm) outer diameter

The failure shown in Fig. 14.2 is the last in a series of four similar failures that were localized in a region above the burners. Note the thick, blunt fracture edge, as well as the shallow gouging along the internal surface. The internal surface was free of deposits.

Microstructural examinations revealed the discontinuous, randomly oriented microcracks directly beneath the corroded regions that are typical of hydrogen damage. Thermal alteration of the microstructure from overheating had not occurred. The iron oxides covering the corroded sites were highly stratified.

Records indicated that condenser in-leakage of low-level chloride salts had induced a depression of boiler-water pH. The presence of the acid-producing salts, coupled with DNB at the rupture site, caused this failure.

CASE HISTORY 14.3

Industry:	Utility
Specimen Location:	Waterwall
Specimen Orientation:	Vertical
Years in Service:	2
Water-Treatment Program:	Coordinated phosphate
Tube Specifications:	2½ in. (6.3 cm) outer diameter

Numerous hydrogen-damage failures caused by caustic corrosion necessitated a major boiler overhaul and significant changes to the water chemistry. A coordinated phosphate program replaced a low-solids, free-alkalinity program. The failure illustrated in Fig. 14.4 occurred in a tube that had been replaced at the time this change in water chemistry occurred.

The failed tube exhibits a thick-walled rupture through a distinct area of gouging on the internal surface. The rupture is immediately downstream of a circumferential weld that protrudes on the internal surface.

Microstructural examinations reveal no thermal alteration of the micro-

Figure 14.4 Rupture resulting from hydrogen damage. Note proximity of failure to circumferential weld. *(Courtesy of Electric Power Research Institute.)*

structure as a consequence of overheating. However, fine discontinuous intergranular microcracks are present in the tube wall adjacent to gouged regions. Gouged regions are covered with thick, stratified layers of dense iron oxide.

Highly localized concentration of a corrosive substance on the metal surface caused gouging and hydrogen damage. Concentration of such substances occurred during DNB, which resulted from disrupted water flow past the protruding circumferential weld. The stratified character of the iron oxides that cover the gouged regions indicates that gouging and hydrogen damage were caused by a low-pH environment that resulted from the concentration of acid-producing salts.

CASE HISTORY 14.4

Industry:	Utility
Specimen Location:	Platen tube
Years in Service:	26
Water-Treatment Program:	Coordinated phosphate
Drum Pressure:	2080 psi (14.3 MPa)
Tube Specifications:	3¼ in. (8.3 cm) outer diameter

The massive failure illustrated in Figs. 6.3 and 6.4 (page 87) was the first tube failure to have occurred at this station. The failure occurred 6 months after an acid cleaning of the boiler.

The thick-walled, longitudinal rupture is confined to the zone of corrosion, while the remainder of the internal surface is smooth and uncorroded.

Microstructural examinations revealed no thermal deterioration of the microstructure. A population of discontinuous intergranular microcracks were present in the tube wall immediately below the corroded zone.

The visual and microstructural appearance of the tube revealed that the failure resulted from hydrogen damage caused by exposure to a low-pH substance. The exact source of the substance and mode of concentration is uncertain. Since platen regions of a boiler may be difficult to rinse following an acid cleaning, it is possible that unneutralized acid remaining beneath deposits reacted with the tube metal during boiler operation. The hydrogen damage that was responsible for this failure occurs only during boiler operation.

CASE HISTORY 14.5

Industry:	Utility
Specimen Location:	Nose slope, waterwall
Specimen Orientation:	45°
Years in Service:	25
Water-Treatment Program:	Coordinated phosphate
Drum Pressure:	2000 psi (13.8 MPa)
Tube Specifications:	3 in. (7.6 cm) outer diameter

The thick-walled rupture shown in Fig. 14.5 is one of numerous similar failures recurring in both the nose-arch section and roof tubes, requiring this area of the boiler to be rebuilt. The boiler is in peaking service.

The rupture coincides with a distinct zone of deep metal loss on the internal surface (Fig. 14.6). The wavelike contour of the corroded region is covered with black iron oxide.

Figure 14.5 Thick-walled rupture.

Examination of the microstructure revealed no thermal alteration. Numerous randomly oriented intergranular microcracks were present in the tube wall just below the corroded region.

Localized DNB resulted in concentration of sodium hydroxide, which caused deep caustic gouging at this site. The hydrogen damage and resulting fracture were a direct consequence of the caustic gouging.

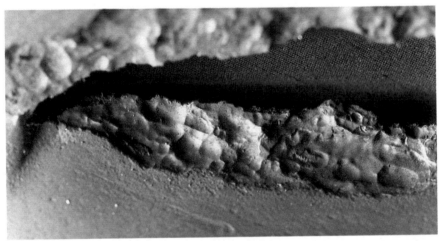

Figure 14.6 Smooth, wavelike contours of internal surface resulting from caustic gouging.

Chapter 15

Corrosion-Fatigue Cracking

Locations

Corrosion fatigue can occur in any location where cyclic stresses of sufficient magnitude are operative. Corrosion-fatigue failures most frequently occur in boilers that are in "peaking" service, used discontinuously, or otherwise operated cyclically. Rapid boiler start-up or shutdown can greatly increase the susceptibility to corrosion fatigue. Some serious corrosion-fatigue problems have been eliminated merely by sufficiently modifying start-up and shutdown rates.

Common locations of corrosion-fatigue cracks include wall tubes, reheater tubes, superheater tubes, economizer tubes, deaerators, and the end of the membrane on waterwall tubing. In addition, corrosion fatigue is common at points of attachment or rigid constraint, such as connections to inlet or outlet headers, tie bars, and buckstays.

Cracks have also been observed at grooves along the internal surfaces of boiler tubes that have been only partly full of water (cracks usually run across the grooves), at points of intermittent steam blanketing within generating tubes, at oxygen pits in waterlines or feedwater lines, in welds at slag pockets or points of incomplete fusion, in soot-blower lines where vibration stresses are developed, and in blowdown lines.

General Description

Corrosion fatigue is a form of deterioration that can occur without concentration of a corrosive substance. The term refers to cracks propagating through a metal as a result of cyclic tensile stresses operating in an environment that is corrosive to the metal. The term and definition above are somewhat misleading in the case of boilers, since normal oxidation of metal to magnetite is sufficient to induce corrosion fatigue in the presence of sufficient cyclic tensile stresses.

Cracks develop according to the following sequence:

- During the first phase of cyclic stress, the tube wall undergoes expansion. Since the oxide layer is brittle relative to the tube wall, the oxide layer may fracture, opening microscopic cracks through the oxide to the metal surface.

Figure 15.1 Incipient corrosion-fatigue crack formed at base of cracked layer of iron oxide. (Magnification: 400X. Etchant: Picral.) *(Courtesy of National Association of Corrosion Engineers.)*

Figure 15.2 Mature corrosion-fatigue crack. (Magnification: 200X. Etchant: Nital.) *(Courtesy of National Association of Corrosion Engineers.)*

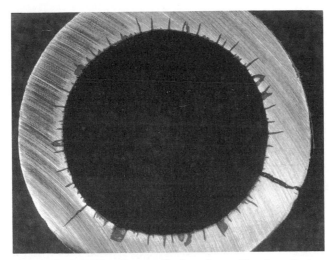

Figure 15.3 A family of longitudinal cracks resulting from fluctuation in internal pressure. *(Courtesy of National Association of Corrosion Engineers.)*

Figure 15.4 Transverse cracks originating on the internal surface.

- The exposed metal surface at the root of the crack oxidizes, forming a microscopic notch in the metal surface (Fig. 15.1).

- During the next expansion cycle, the oxide will tend to fracture along this notch, causing it to deepen.

- As this cyclic process continues, a wedge-shaped crack propagates through the tube wall (Fig. 15.2), until rupture occurs or the tube wall is penetrated.

The cracks always propagate in a direction perpendicular to the direction of the principal stress. Hence, if the principal cyclic stress is produced by fluctuations in internal pressure, longitudinal cracks are produced (Fig. 15.3). If the principal cyclic stress is a bending stress produced by thermal expansion and contraction of the tube, cracks will be transverse (Fig. 15.4). Corrosion-fatigue cracking commonly occurs adjacent to physical restraints. Cracks may originate on the external surface, the internal surface, or both simultaneously. Cracks originating on the internal surfaces are often associated with pits. The pit site serves as a stress-concentrating notch, making it a preferred site for initiation of corrosion-fatigue cracks.

Critical Factors

Cyclic tensile stresses and an environment that will cause spontaneous oxidation of a bare metal surface are two critical factors that govern susceptibility to corrosion fatigue. Two common sources of cyclic tensile stresses are cyclically fluctuating internal pressure, and constrained thermal expansion and contraction.

In addition to an environment that will cause spontaneous oxidation of a bare metal surface, other factors that may provide a significant contribution are pH level and dissolved-oxygen content. Operation at low-pH levels or with excessively high levels of dissolved oxygen may induce pitting. The pits act as stress concentrators for the initiation of corrosion-fatigue cracks.

Identification

Corrosion-fatigue cracks are typically straight and unbranched. They are needle- or wedge-shaped, and propagate perpendicularly to the metal surface. They often occur in families of parallel cracks (Figs. 15.3 and 15.4), and are frequently very tight, making them difficult to see without very close examination. At times, they may appear to be only shallow grooves in the magnetite covering. Typically, they do not run long distances along the tube surface. It is not unusual for corrosion-fatigue cracks to develop simultaneously within two or more components of similar location.

Nondestructive methods for crack identification include ultrasonic surveillance, radiographic surveillance, dye penetrant, and magnetic-particle inspection.

Elimination

Elimination or reduction of corrosion-fatigue cracking is realized by controlling cyclic tensile stresses, controlling environmental factors, and boiler redesign. Reducing or eliminating cyclic operation of the boiler as well as extending start-up and shutdown times may help eliminate or reduce corrosion-fatigue cracking.

Elimination of the oxidation process occurring at a newly exposed crack tip is not feasible. This oxidation will occur spontaneously even at very low levels of dissolved oxygen. Controlling pH and excessive levels of dissolved oxygen can be useful in eliminating pitting corrosion, which will eliminate a common point of initiation for corrosion-fatigue cracking.

In persistent cases of corrosion-fatigue cracking, measures such as contouring of welds and redesign of tube attachments may be required to eliminate or reduce constraints to thermal expansion and contraction.

Cautions

Complete fractures resulting from corrosion-fatigue cracking are typically thick-walled and show very little, if any, ductility. These fractures might conceivably be confused with other failure modes that typically produce thick-walled fractures, such as stress rupture, cracking caused by hydrogen damage, stress-corrosion cracking, and some types of severe overheating. Corrosion-fatigue cracks are frequently difficult to see since they are often filled with dense iron oxides. At times they may appear as short, shallow grooves in the magnetite layer covering the tube.

Related Problems

See also Chap. 2, "Long-Term Overheating"; Chap. 3, "Short-Term Overheating"; Chap. 14, "Hydrogen Damage"; and Chap. 16, "Stress-Corrosion Cracking."

CASE HISTORY 15.1

Industry:	Utility
Specimen Location:	Cold reheat line 20 in. (50.8 cm) from turbine discharge
Specimen Orientation:	Horizontal
Drum Pressure:	350 psi (2.4 MPa)
Tube Specifications:	14 in. (35.6 cm) outer diameter, seamless

The cracks illustrated in Fig. 15.4 were confined to zones 5 to 10 ft (1.5 to 3.0 m) on either side of a coupling in the reheat line. Failures had not occurred, but similar cracking had been observed in a sister unit previously. The transverse fissures and cracks were accompanied by, and associated with, a population of small pits (Fig. 15.5). Reddish iron oxides were present on this surface.

Microstructural examinations revealed families of blunt, V-shaped fissures entering the wall from the internal surface. The deepest fissures penetrated 25% of the wall thickness.

Figure 15.5 Transverse cracks and associated oxygen pits.

The cracking in this case was due to cyclic flexing of the line. Small oxygen pits, resulting from exposure of the internal surfaces to moisture and oxygen during idle times, served as nucleation sites for the corrosion-fatigue cracks.

Mitigation of this problem can be achieved by eliminating oxygen pitting during idle periods, and by reducing or eliminating the cyclic flexing of the line.

CASE HISTORY 15.2

Industry:	Utility
Specimen Location:	Drain line off reheat header
Tube Specifications:	1⅞ in. (4.8 cm) outer diameter, low-alloy steel

Deep, cross-hatched cracks and fissures are located in a distinct zone around one end of the internal surface of the drain line (Figs. 15.6 and 15.7). Cracks and fissures are not present in areas away from this end.

Microstructural examinations revealed classic thermal-fatigue cracks propagating through the wall. Evidence of mild overheating was also observed.

Figure 15.6 Thermal-fatigue cracks on internal surface.

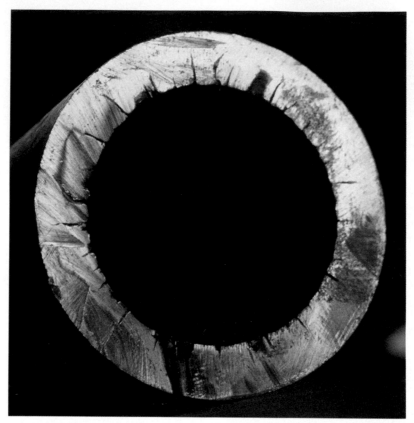

Figure 15.7 Cross section through cracks.

Thermal fatigue can occur when heated metal is rapidly cooled repeatedly. Rapid cooling of the metal surface establishes high triaxial stresses that can produce cross-hatched cracks of the type illustrated in these figures. Rapid cooling may have been induced by localized exposure of the line to slugs of water. Elimination of the rapid cooling cycles is necessary to eliminate this problem.

CASE HISTORY 15.3

Industry:	Pulp and paper
Specimen Location:	Front-wall tube around primary air port, recovery boiler
Specimen Orientation:	Slanted
Years in Service:	15
Drum Pressure:	900 psi (6.2 MPa)
Tube Specifications:	3 in. (7.6 cm) outer diameter, studded

The cracking illustrated in Figs. 15.8 and 15.9 was localized to the area around the primary air port. These short transverse fissures were especially prominent near the base of studs (Fig. 15.9). Measurement revealed penetrations of 15 to 20% of the original tube-wall thickness.

Microstructural examinations of surface profiles revealed deep, wedge-shaped fissures from the external surface and shallow, sharp cracks from the internal surface.

The transverse orientation of cracks and fissures reveals that they were produced by cyclic, outward bending of the tubes resulting from thermal expansion and contraction. The prominence of the cracks at the stud bases may be due to differences in the thermal expansion and contraction charac-

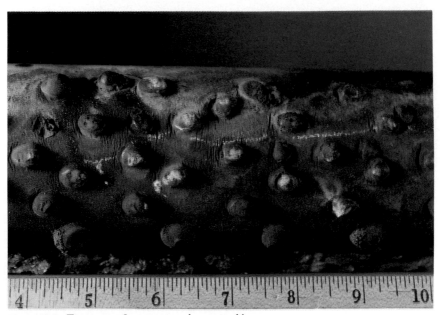

Figure 15.8 Transverse fissures at and near stud bases.

teristics of the studs and the tube wall. The visual appearance of such corrosion-fatigue fissures on studded tubes has led to the term "elephant hiding," to describe the phenomenon. Experience suggests that elephant hiding occurs in areas of high heat-transfer rate, and may occur on tubes that are not covered with smelt.

Figure 15.9 Transverse fissures near stud (center). (Magnification: 6.5×.)

CASE HISTORY 15.4

Industry:	Pulp and paper
Specimen Location:	Superheater near outlet header, power boiler
Specimen Orientation:	Vertical
Years in Service:	5
Water-Treatment Program:	Coordinated phosphate
Drum Pressure:	1250 psi (8.6 MPa)
Tube Specifications:	1¾ in. (4.4 cm) outer diameter, low-alloy steel

The thick-walled circumferential fracture shown in Fig. 15.10 was the first superheater failure in this boiler. Close visual examinations of both internal and external surfaces adjacent to the fracture revealed secondary cracks (Fig.

Figure 15.10 Brittle fracture face resulting from corrosion fatigue.

15.11). Microstructural examinations of the tube wall confirmed the presence of families of unbranched transgranular cracks near the fracture both internally and externally. The circumferential orientation of the cracks reveals that the stresses responsible were cyclic bending stresses, possibly caused by thermal expansion and contraction of the tube.

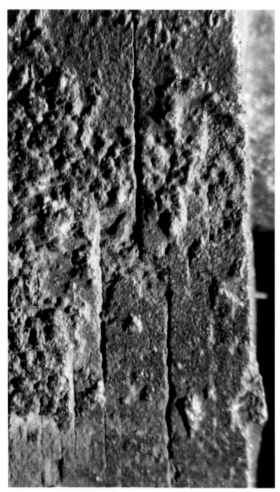

Figure 15.11 Secondary corrosion-fatigue cracks on external surface. (Magnification: 6.5×.)

Chapter 16

Stress-Corrosion Cracking

Locations

In principle, stress-corrosion cracking could occur wherever a specific corrodent and sufficient tensile stresses coexist. Because of improved water-treatment programs and improved boiler design, the occurrence of caustic stress-corrosion cracking* is much less frequent now than it was years ago. However, stress-corrosion cracking continues to appear occasionally in water tubes, superheater tubes, and reheater tubes. Stress-corrosion cracking may also occur in stressed components in the steam drum, such as bolts (Fig. 16.1).

General Description

The term *stress-corrosion cracking* refers to metal failure resulting from a synergistic interaction of a tensile stress and a specific corrodent to which the metal is sensitive. The tensile stresses may be either applied, such as those caused by internal pressure, or residual, such as those induced by

---

Caustic embrittlement is presently considered an obsolete, historical term for this phenomenon.

Figure 16.1 Steam drum bolt that failed by caustic stress-corrosion cracking. Note the typical brittle character of the fracture. *(Courtesy of National Association of Corrosion Engineers.)*

welding. In boiler systems, carbon steel is specifically sensitive to concentrated sodium hydroxide, while stainless steel is specifically sensitive both to concentrated sodium hydroxide and to chlorides.

Gross attack of the metal is not necessary for this phenomenon and, in fact, does not characteristically accompany it. The combination of concentrated sodium hydroxide, some soluble silica, and tensile stresses will cause continuous intergranular cracks to form in carbon steel (Fig. 16.2). As the cracks progress, the strength of the remaining intact metal is exceeded, and a brittle, thick-walled fracture will occur.

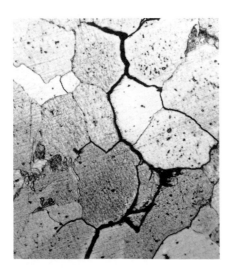

Figure 16.2 Caustic stress-corrosion cracking: continuous intergranular cracks running through a tube wall. (Magnification: 500✕.)

Critical Factors

There are two principal factors that govern stress-corrosion cracking in the boiler environment. First, the metal in the affected region must be stressed in tension to a sufficiently high level. The stresses may be applied and/or residual. Second, concentration of a specific corrodent at the stressed metal site must occur. The specific corrodent for carbon steels in boiler systems is sodium hydroxide; for stainless steels the corrodent can be sodium hydroxide or chlorides. (Concentration of corrodents may occur by any of the modes outlined in Chap. 4, "Caustic Corrosion.") Small leaks may also result in concentration of corrodents.

Instances of caustic stress-corrosion cracking in boiler metal operating below 300°F (149°C) are rare. Concentrations of sodium hydroxide as low as 5% have caused cracking, but concentrations in the range of 20 to 40% greatly increase susceptibility.

Identification

Failures caused by stress-corrosion cracking always produce thick-walled fracture faces regardless of the degree of metal ductility. Branching is frequently associated with these cracks. Unless failure has occurred, stress-corrosion cracking may be difficult to see with the unaided eye, since the cracks tend to be very fine and tight. Occasionally, evidence of the presence of concentrated sodium hydroxide, such as whitish, highly alkaline deposits or the presence of crystalline magnetite, may be observed at the crack site.

Elimination

To eliminate problems with stress-corrosion cracking it is necessary to gain control of either tensile stresses or concentration of corrodents.

Tensile stresses can be either applied or residual. *Applied stresses* are service-generated stresses including hoop stresses caused by internal pressure and bending stresses from constrained thermal expansion and contraction. Generally, hoop stresses are subject to minimal control, since the essential function of the boiler tubes and other pressurized components demands containment of pressurized substances. Bending stresses, however, may be reduced or eliminated by altering operational parameters or by redesign of the affected components.

The term *residual stress* refers to stresses that are inherent in the metal itself. They are the result of manufacturing or construction processes such as welding or tube bending. Residual hoop stresses may also remain from the manufacturing process. These stresses can be relieved by proper annealing techniques.

Avoiding concentrated corrodents is generally the most successful means of reducing or eliminating stress-corrosion cracking. Avoiding departure from nucleate boiling (DNB), keeping internal surfaces sufficiently free of deposits, and avoiding formation of steamlines and waterlines in components receiving high heat flux, are first steps. Other steps may include preventing in-leakage of alkaline-producing salts through condensers, heat exchangers, process streams, and caustically regenerated demineralizer systems; preventing contamination of desuperheating or attemperator water by alkaline materials or chlorides; and preventing boiler-water carryover.

The use of inhibitors, such as sodium nitrate or a combination of sodium nitrate and one of many selected organics, has been successful in reducing caustic stress-corrosion cracking. A coordinated phosphate program, which is designed to eliminate the formation of free sodium hydroxide, may also be valuable.

Cautions

Stress-corrosion cracks are commonly difficult to identify through visual inspection. The use of dye penetrants, magnetic-particle inspections, and ultrasonic testing in suspect regions may disclose the presence of stress-corrosion cracks. Dwell time for dye penetrants must be increased to accommodate the typical tightness of these cracks.

It is conceivable that stress-corrosion cracking could be confused with other cracking modes that produce thick-walled fractures, such as hydrogen damage, corrosion fatigue, creep rupture, and some forms of severe overheating. Confirmation of a diagnosis of stress-corrosion cracking requires metallographic examination.

Related Problems

See also the section titled Creep Rupture in Chap. 2, "Long-Term Overheating"; Chap. 3, "Short-Term Overheating"; Chap. 14, "Hydrogen Damage"; and Chap. 15, "Corrosion-Fatigue Cracking."

CASE HISTORY 16.1

Industry:	Pulp and paper
Specimen Location:	Steam supply line to a soot blower, recovery boiler
Specimen Orientation:	Vertical
Years in Service:	15
Water-Treatment Program:	Oxygen scavenger
Drum Pressure:	400 psi (2.8 MPa)
Tube Specifications:	2⅜ in. (6.0 cm) outer diameter, 304 stainless steel

Figure 16.3 shows one of several stainless steel soot-blower lines that had cracked. Failures of this type can be quite dangerous, since steam can be released into occupied areas.

The line contained several thick-walled cracks that ran as long as 25 in. (63.5 cm) down the tube. Note that the cracks are not tight but have spread apart, an unusual feature for stress-corrosion cracking.

Close visual and microstructural examinations revealed that the cracks originated on the external surface. Microstructural examinations also revealed severe cold working of the metal.

An uncontaminated segment of the crack face was examined under a scanning electron microscope equipped for energy-dispersive spectroscopy.

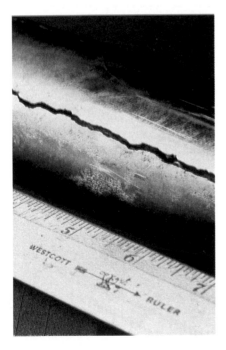

Figure 16.3 Extensive longitudinal crack in a stainless steel soot-blower line. Note that the crack edges have spread apart.

Elemental analysis of substances covering the crack face revealed the presence of chlorine.

The line failed by stress-corrosion cracking resulting from exposure of the external surface to chlorides of unknown origin. The longitudinal direction of the cracking reveals that circumferential (hoop) stresses were responsible. It is apparent from the spreading apart of the cracks and the evidence of cold-worked metal in the microstructure that the tube contained high residual hoop stresses resulting from tube-forming processes. The level of residual stresses was calculated to be 29,000 psi (200 MPa). The residual stresses, added to service stresses generated by internal pressure, acted synergistically with the chlorides to produce the cracks illustrated.

CASE HISTORY 16.2

Industry:	Pulp and paper
Specimen Location:	Economizer
Specimen Orientation:	Bend
Years in Service:	15
Water-Treatment Program:	Phosphate
Drum Pressure:	800 psi (5.5 MPa)
Tube Specifications:	2 in. (5.1 cm) outer diameter

A section of an economizer tube having a 90° bend at its midpoint contains a pair of thick-walled cracks on opposite sides of the tube (Fig. 16.4). The tube had an oval rather than a circular cross section through the bend, and the cracks were located at opposite ends of the long axis of the oval.

The internal surface had sustained shallow, general metal loss. Sparkling crystals of black magnetite were associated with this corrosion.

Microstructural examinations revealed that the cracks originated at the bottom of shallow pits. The cracks exhibited branching as they propagated through the tube wall. Crack paths were principally across the metal grains (transgranular).

Stress-corrosion cracking in carbon steels requires the joint action of concentrated sodium hydroxide and tensile stresses. The sodium hydroxide apparently concentrated beneath porous iron oxide deposits that were present on the internal surface. The presence of crystalline magnetite indicates exposure to concentrated sodium hydroxide. The shallow pits acted as stress concentrators, elevating the normal stress level. In addition, stresses from internal pressure would be highest along the narrow ends of the oval cross section, where the cracks formed. The longitudinal orientation of the cracks reveals that internal pressure provided the stresses necessary for stress-corrosion cracking.

Figure 16.4 Longitudinal crack along the side of a 90° bend in an economizer tube. A similar crack is located along the opposite side.

CASE HISTORY **16.3**

Industry:	Pulp and paper
Specimen Location:	Superheater, recovery boiler
Specimen Orientation:	Vertical
Years in Service:	11
Water-Treatment Program:	Coordinated phosphate
Drum Pressure:	1200 psi (8.3 MPa)
Tube Specifications:	2 in. (5.1 cm) outer diameter, 304 stainless steel
Fuel:	Black liquor

Recurrent failures of the type illustrated in Figs. 16.5 and 16.6 plagued the superheater section of this boiler. Note the fine cracks adjacent and parallel to the weld. Surface branching of the crack is apparent in Fig. 16.6.

Microstructural examinations revealed branched cracks running across the grains (transgranular) and originating on the internal surface. The cracks are located in the heat-affected zone immediately adjacent to the weld.

This failure is caustic stress-corrosion cracking. The source of sodium hydroxide is probably carryover from the steam drum. Stresses are residual welding stresses, as indicated by the circumferential crack propagation and the proximity of the cracks to the weld. Cracks resulting simply from stresses imposed by internal pressure would be longitudinally oriented.

Figure 16.5 Appearance of stress-corrosion crack on the external surface. Crack is located within the red circle.

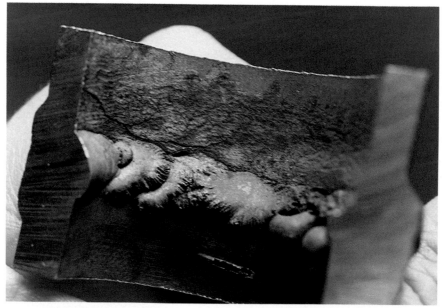

Figure 16.6 Appearance of crack on the internal surface. Note the branching and proximity to the weld bead.

CASE HISTORY 16.4

Industry:	Petrochemical
Specimen Location:	Superheater, first stage
Specimen Orientation:	Vertical
Years in Service:	3 weeks
Water-Treatment Program:	Phosphate
Drum Pressure:	600 psi (4.1 MPa)
Tube Specifications:	1½ in. (3.8 cm) outer diameter, 304 stainless steel
Fuel:	Waste gas

The tube illustrated in Fig. 16.7 is one of numerous tubes that failed in this boiler. The tubes had been moderately cold-bent during installation, and were not stress-relief-annealed. The steam drum lacked adequate devices for separation of steam and water, and load swings were frequent, possibly causing carryover of boiler water.

Microstructural analysis revealed plastically deformed grains from the cold bending. The cracks were highly branched and ran between the grains (intergranular) as they passed through the tube wall. They originated on the internal surface.

The source of stress in this case was residual stresses from the bending operation. This is apparent from the circumferential orientation of the cracks, the fact that the cracks have spread apart, and the proximity of the cracks to the bent zones of the tubes. The corrodent was sodium hydroxide from boiler-water carryover.

This case is an excellent example of the value of understanding failure mechanisms and performance of materials. Stainless steel superheater tubes in a boiler of this pressure are unusual. The original tubes were carbon steel that cracked after 9 months of service. Two steps were taken to improve service life. First, stainless steel tubes were specified to replace the carbon steel. Second, moderate bends were put in the tubes, apparently to relieve the thermal expansion and contraction stresses that had caused cracking in the carbon steel tubes. Unfortunately, despite the greater general corrosion resistance of stainless steels, they are also susceptible to caustic stress-corrosion cracking. In addition, the bends placed in the tubes to relieve thermal stresses provided high residual stresses instead. The stainless steel tubes failed after 3 weeks.

Figure 16.7 Transverse crack resulting from caustic stress corrosion in a bent stainless steel superheater tube. Note small "window" that has been blown out of the wall.

Chapter 1

Erosion

Introduction

The four major locations in which boiler erosion occurs are at the fire side, the preboiler, the afterboiler, and the water side and steam side.

Fire-side mechanisms cause most erosion-related failures and can be further subdivided as erosion related to soot blowing, steam cutting, fly-ash attack, coal-particle impingement, and falling slag. Fluidized-bed and other special-purpose boilers sometimes suffer severe attack. Boilers burning wood chips and bagasse are often eroded by entrainment of tramp contaminants such as sand and other foreign material in furnace gases. Incinerators suffer similar fire-side problems. Because fire-side mechanisms cause most erosion-related failures, each mechanism will be discussed in detail.

Erosive metal loss on water-side surfaces is comparatively rare. Cases do occur, however. Internal-surface discontinuities or solid foreign objects lodged within tubes can disturb flow, increase turbulence, and cause wastage.

Preboiler attack is confined primarily to feedwater systems. Turbine erosion is common in afterboiler regions. Burner nozzles, blowdown piping, condensate return lines, and many other boiler components are also eroded.

Soot-Blower Erosion

Locations

As the term *soot-blower erosion* implies, damage occurs near or in the direct path of soot-blower discharge. Superheater tubing is usually attacked. Common damage locations include tubes along the path of retractable soot

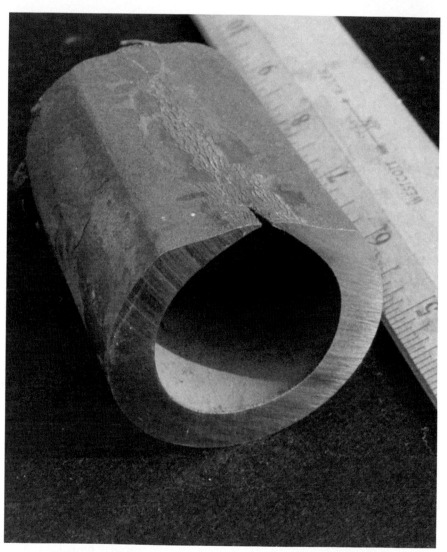

Figure 17.1 Superheater section thinned by soot-blower erosion. Note the flattened surface. The tube bulged and ruptured on the thinned side. Also, note how the eroded surface color is not dissimilar to unattacked regions, indicating intermittent attack.

blowers, and particularly those tubes nearest wall entrances of retractable blowers. Other damage locations include furnace corners opposite wall blowers. Platens in the convection section are often targets, as are any tubes near malfunctioning soot blowers.

General description

Perhaps the most common cause of erosion in boilers is soot-blower attack. Usually a misdirected blower allows a high-velocity jet of steam or air carrying condensed water droplets to impinge directly upon tube surfaces, rather than to be directed between tubes. Physical abrasion and accelerated oxidation cause metal loss. Damage can be accelerated by fly ash entrained in the high-velocity jet stream directed against the tube surface. Erosive thinning often leads to tube rupture (Figs. 17.1 and 17.2).

Critical factors

Soot-blower erosion is caused by improper blower alignment. Entrainment of either condensed water or fly ash in the blower gas also accelerates attack. Raising blowing pressure increases gas velocity and thus promotes damage by entrained fly ash. Improper alignment and operation are the most common sources of damage.

Figure 17.2 Tube rupture caused by a misdirected soot blower. Rupture edges are thin and ragged because of tearing of the wasted steel in the eroded zone. The tube surface is light gray and shiny as a result of erosion of the normally present fire-side oxides and deposits.

Identification

Attacked surfaces are locally thinned, usually producing longitudinally aligned zones of flattened metal bordered on both sides by shoulders of unattacked metal. When viewed in transverse cross section, the tube appears to have been "planed" along its length (Figs. 17.1 and 17.2). Grooving and more irregular attack will be present if eddies and gas channeling are pronounced.

Metal surfaces will be smooth or have smoothly undulating contours. Only a thin, dark, oxide layer, or no oxide layer at all, will be present if attack is fresh. If attack is old or intermittent (as is often the case with soot-blower attack), oxide and deposits will cover thinned surfaces. However, the oxide and deposit layers will usually be thinner than on adjacent unattacked surfaces. Close visual observation of thinned surfaces will often show very shallow wavelike striations frozen into the metal surface (Figs. 17.3 and 17.4). The striations will be aligned perpendicularly to steam flow across the surface.

Attack usually is present at the opening of the soot-blower valve and continues along the blower path, decreasing in severity as the distance

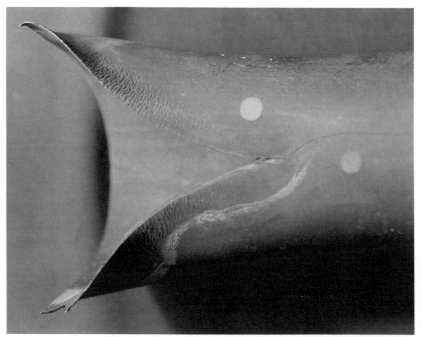

Figure 17.3 Numerous parallel surface ripples at the edges of an external-surface rupture in a superheater tube. The light circles are regions where acid drops removed the thin oxide layer during laboratory testing.

Figure 17.4   Finely spaced, wavelike, parallel undulations on eroded fire-side external surface of superheater tube. The striations resemble waves viewed from the air and are aligned perpendicularly to the direction of steam impingement. (Compare with Fig. 17.3.)

from the blower increases. Frequently, the blower is misaligned, is frozen in position, or is blowing wet steam because of water entrainment.

Bulging will usually be absent in internally pressurized tubes suffering erosion. Rupture is usually longitudinal but can follow grooves whatever their orientation. Rupture edges will be thin and ragged.

Elimination

Proper soot-blower operation, alignment, and function are required to reduce attack. Periodic inspection of nozzle position and alignment is recommended. Elimination of moisture in blowing steam can be accomplished by allowing adequate steam warm-up and providing for adequate drainage of steam supply lines. Metallizing, welding, and plasma spray techniques can increase wall thickness in affected areas and thereby prolong tube life. Coating processes, however, can slightly reduce heat transfer locally and do not remove the root problem.

Cautions

Soot-blower damage can resemble oil-ash or coal-ash corrosion, as well as fly-ash erosion. If heavy deposit accumulations are present atop thinned tube walls, damage may be due to oil-ash or coal-ash corrosion, even though these tubes exhibit other characteristics of erosion. Boilers burning wood, bagasse, and other waste materials are particularly likely to suffer erosion since sand, dirt, and cement dust may be entrained in combustion gases. Attack by entrained particulate resembles soot-blower attack. Location of damage in the direct path of the soot-blower stream is necessary for diagnosis of failure.

Related problems

See also Chap. 9, "Oil-Ash Corrosion"; Chap. 10, "Coal-Ash Corrosion"; and the sections titled Steam Cutting from Adjacent Tube Failures, Fly-Ash Erosion, Coal-Particle Erosion, and Falling-Slag Erosion in this chapter.

Steam Cutting from Adjacent Tube Failures

Locations

Affected tubes can occur in any part of the boiler. Cutting is most severe when internal steam pressures, and consequently temperatures, are high. Thus, superheaters and reheaters are often severely attacked. A nearby failed tube is always present. Usually damage is highly localized and is worst in line of sight with the nearby failed tubes. Occasionally, if the original failure is not detected promptly, a single failure leads to chain reactions involving multiple tube breaches.

General description

Damage produced by escaping high-velocity fluids eventually causes other nearby tubes to be steam-cut. The wastage mechanism is essentially the same as soot-blower attack. However, attack is usually more localized. Wastage occurs rapidly.

Critical factors

Pressure of escaping fluids and proximity of affected tubes dictate the damage potential. Wastage rates increase as pressures (and consequently temperatures) increase, and distance decreases.

Identification

Wasted surfaces sometimes resemble those produced by soot-blower attack (see the section titled Soot-Blower Erosion). Surfaces close to nearby leaking tubes are likely to be smoothly undulating and grooved (Figs. 17.5 and 17.6). The irregularity of wastage increases as distance to the erosion source decreases. Freshly attacked surfaces contain almost no oxide or deposition. Surfaces are usually shallowly pebbled or striated and have an undulating contour. Often, the damage can be directly correlated with the nearby failure by line-of-sight extrapolation.

Elimination

Since steam cutting is caused by other unpredictable tube failures, and such failures can occur anywhere in the boiler, it is not practical to design or plan for the damage. The only reasonable way to reduce the incidence of failure is to reduce the likelihood of other failures.

Cautions

Steam cutting is usually obvious because of associated failure or failures. However, such damage can sometimes be confused with damage from other forms of fire-side erosion including fly-ash, coal-particle and falling-slag erosion. Also, once the steam-cut tube begins leaking, escaping fluids from

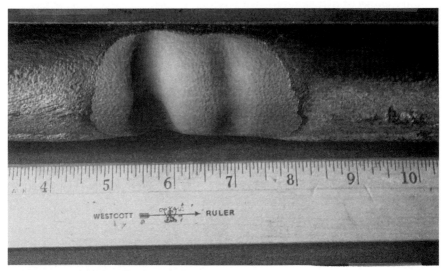

Figure 17.5 Utility superheater tube cut by steam leaking from an adjacent tube that failed at a manufacturing defect. Note the smoothly undulating surface and removal of red and black oxides and deposits in the wasted area.

Figure 17.6 Grooving on pendant U-bend superheater leg caused by steam leaking through a corrosion-fatigue crack at the U-bend. Grooves were rusted after removal from boiler.

subsequently damaged tubes can modify the nearby failure that was the primary cause of attack (see Case History 17.10).

Related problems

See also the sections titled Soot-Blower Erosion, Fly-Ash Erosion, Coal-Particle Erosion, and Falling-Slag Erosion in this chapter.

Fly-Ash Erosion

Locations

Damage frequently occurs in economizer, superheater, reheater, and roof tubing, although other tubes may be affected. Since fly ash is usually more erosive when particle temperatures are lower, economizers are frequently attacked. Inlet areas of reheaters are common wastage sites because of higher gas velocities and eddying present there. Areas in the superheater where slagging is pronounced are common problem regions, as gas flow in the narrow slag channels increases. Any location where channeling or eddying of gases occurs is susceptible to wastage. Erosion is usually localized and frequently is restricted to regions such as gaps between tube rows, banks, and duct walls.

General description

Fly-ash erosion is caused by particulate matter entrained in high-speed flue gases striking metal surfaces. Major accelerating factors are high gas velocity and large amounts of abrasive components in the fly ash. These factors accelerate loss by increasing the amount of kinetic energy per impact and by increasing the number of impacts per unit time in a given area. Table 17.1 shows maximum design gas velocities for various types of firing or fuel.

Fly-ash erosion is common in boilers fired with overfeed stokers, which allow considerable amounts of ash to enter the gas stream more readily. Those boilers using an overfire air system may reduce particulate. Partial-suspension burning causes greater amounts of particulate matter to enter the gas stream. Thus, collectors are generally used. When collectors malfunction, damage increases.

TABLE 17.1   Design Gas Velocity (fps) through Net Free-Flow Area in Tube Banks to Prevent Flue-Dust Erosion

	Baffle arrangement	
Type of firing or fuel	Multipass	Single pass
Pulverized coal	75	75*
Spreader stoker	50	60
Chain-grate stoker, anthracite	60	75
Chain-grate stoker, coke breeze	60	75
Chain-grate stoker, bituminous	100	100
Underfeed stoker	75	100
Blast-furnace gas	75	100
Cyclone furnace	——	100
Wood or other waste fuels containing:		
Sand	50	60
Cement dust	——	45
Bagasse	60	75

* For PC units burning fuels having more than 30% ash on a dry basis, limit the maximum velocity through the free-flow area to 65 fps. For PC units burning coals producing fly ash with known high-abrasive tendencies, such as Korean or central Indian coals, limit the maximum velocity through free-flow area to 45 fps.

NOTE 1 fps = 1 ft/s = 0.3048 m/s

SOURCE: Courtesy Babcock & Wilcox, *Steam/Its Generation and Use*, Babcock & Wilcox, New York, 1972.

Boilers using fuel contaminated with sand and dirt—such as wood chips—suffer damage almost identical to fly-ash erosion.

Critical factors

Erosive metal loss increases as particle hardness, flow velocity, and ash concentration increase. Of these factors, flow velocity and ash concentration are most important (Fig. 17.7). Erosive loss increases as the angle of impingement between gas flow and the metal surface increases. The direct incidence of fly ash is more damaging than glancing blows. As temperature increases, erosive metal loss decreases because particles become softer.

Size, hardness, and composition of particulate matter also influence attack. Particles larger than about 0.001 in. (0.0025 cm) and those containing high concentrations of aluminum and silicon compounds are more erosive because of high particle hardness and large particle kinetic energy.

Identification

Fly-ash erosion frequently causes smoothly polished surfaces. In other cases, irregular flow marks and grooving are produced by eddies around slag encrustations, hangers, brackets, etc. (Fig. 17.8). In extreme cases, thinning can cause rupture. Attack may be localized or general.

Discrimination between fly-ash erosion and other forms of fire-side erosion requires more information.

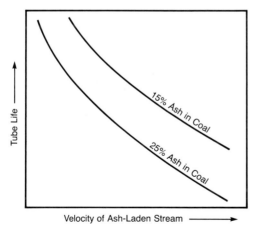

Figure 17.7 Effects of velocity and ash content on the life of a tube. Increases in the number and velocity of the impacting ash particles can result in reduced service life. (*Courtesy of Electric Power Research Institute,* EPRI Manual for Investigation and Correction of Boiler Tube Failures, *EPRI, 1985. Originally in E. Raask, "Tube Erosion by Ash Impaction,"* WEAR, **13**:(1969), pp. 301–315.)

Figure 17.8 Deep grooves cut into the hot side of a waterwall tube. Grooving was caused by channeling of furnace gases containing entrained fly ash.

Locations in the boiler in which attack occurs, positions of deflecting baffles, amount of entrained fly ash, and local gas velocity are important clues to attack. Fire-side oxide and deposit layers are usually much thinner on wasted regions, allowing discrimination from coal-ash or oil-ash corrosion, in which deposits and corrosion-product accumulations are extensive. No chemical attack is present as might be expected in cold-end corrosion. When ruptures occur, they are usually thin-edged. Close inspection of the rupture site often reveals thinning of the external surface and no overheating or corrosion. Metallographic examination may reveal microscopic plastic deformation on thinned surfaces caused by impingement of high-velocity particles.

Elimination

Decreasing fly-ash erosion requires a system operations approach. This includes making sure all baffles, collectors, refractories, and the like are working properly. In extreme cases, redesign of boiler components may be required. Of course, reducing the amount and velocity of fly ash will also limit damage. In extreme cases, fuels less prone to produce erosive ash may have to be used. High load and excess air increase flue-gas velocity, and thus, increase attack.

Slagging promotes fly-ash erosion by channeling gases and increasing eddying. Appropriate fuel additives and soot blowing can reduce slagging. Baffles have been used to distribute flue gases, and consequently, fly ash, more evenly. However, where gas flow is horizontal through tube banks, baffling is generally absent. Slag fences have been used to prevent larger slag pieces from entering horizontal tube banks. Shielding and metal spray coatings are beneficially used in certain erosion-prone locations.

Cautions

Fly-ash erosion resembles other forms of erosive attack, including soot-blower erosion, impingement from nearby leaking tubes, coal-particle erosion, falling-slag erosion, and coal-ash and oil-ash attack. Cold-end corrosion also can be confused with erosion by fly ash. Location of failures and knowledge of boiler operation are important in making a correct diagnosis.

Related problems

See Cautions above.

Coal-Particle Erosion

Locations

Water-cooled tubes that line the cylindrical combustion chamber of cyclone-type coal burners are common attack sites. Incompletely burned coal particles can also accelerate fly-ash erosion in superheaters and wall tubing. Attack is common in utility boilers.

General description

In general, attack occurs in tubes lining the cyclone burner. Erosion becomes pronounced when the refractory covering the tubes or the wear liners is damaged or worn out, exposing unprotected tubes. The high-velocity coal particles, moving at speeds up to 300 ft/s (9.0 m/s) (in utility service), impinge on tube walls and cause rapid wear. Attack is similar in appearance to fly-ash erosion.

Critical factors

The degradation of the refractory and wear liners contributes to attack. Low-silicate coals produce less damage.

Identification

In most cases, diagnosis of coal-particle erosion is simple. Damage closely resembles fly-ash erosion but occurs at or near a cyclone burner and damages the refractory or wear liners. Inspection usually reveals damaged areas and the associated spalled refractory.

Elimination

Frequent inspection and periodic maintenance of the refractory and wear liners will eliminate most damage.

Cautions

Not all attacked tubes need to be replaced. However, tubes should always be inspected to determine wall thickness after they are found to be damaged.

Related problems

See the discussion under Fly-Ash Erosion in this chapter.

Falling-Slag Erosion

Locations

This damage is rare and usually is confined to slanted tube walls near the bottom of large boilers, which direct ash into the ash hopper. Most damage occurs near side walls, where greater amounts of slag tend to accumulate since slag from side walls is more likely to strike these areas.

General description

Erosion is caused by slag particles that fall from above. The particles strike the slanted walls and cause wastage. If large slag particles fall, they can dent and bend tubes.

Critical factors

The amount of slag per unit area, and the tendency of this slag to fall, control the damage rate. Slag formation is favored in a furnace containing a large wall area; having a low or high flue-gas velocity; burning a high-slagging coal containing high sodium and chlorine concentrations; having low flue-gas temperature; or cycling thermally and/or running high tube temperatures. Such slag can be easily shed.

Identification

Falling-slag erosion produces flat spots on slanted tubes. If large slag chunks falling from great heights strike surfaces, tubes can be dented and deformed. Microscopic examination often reveals the plastic deformation associated with this damage.

Elimination

Reduction of slagging decreases damage. This may necessitate a change of coal type. The higher the fusion temperature of the coke ash, the lower the slagging rate. Use of fuel-additive chemicals may also reduce slagging potential. Structural alterations such as increasing tube-wall thickness, and use of weld overlays or wear bars have been used to extend tube life.

Cautions

Falling-slag erosion can resemble erosion caused by fly ash, coal dust, soot blowers, and steam cutting. However, the widespread attack on slanted tubes near ash hoppers is usually definitive.

Related problems

See the discussion under Fly-Ash Erosion in this chapter.

Preboiler and Afterboiler Erosion

Locations

Feedwater-pump components, including impellers, fittings, valves, and housings are often eroded. Less commonly, transfer lines, pipe elbows, and blowdown components are attacked.

Afterboiler erosion is confined primarily to turbines. Because of high velocities inherent in turbine operation, turbine components frequently suffer erosion. Turbine blades are wasted both by hard particles and by water-droplet impingement. Latter-stage buckets are most frequently affected by water-droplet impingement. Carbon steel piping may be attacked in the presence of wet, high-velocity steam, causing damage that falls more into the category of erosion-corrosion. Valve stems, nozzle blocks, diaphragms, and early-stage buckets commonly suffer hard-particle erosion due to exfoliated oxide particles from superheaters, reheaters, main steam leads, and hot reheat piping.

General description

Erosion is metal loss caused by impact of solids or liquids. Attack is promoted by turbulent, high-velocity fluid flow. Rapid pressure changes promote water jetting and turbulence. Abrupt changes in flow direction and the entrainment of hard particulate matter in fluids also contribute to wastage. Although appearing simple, the erosion process is complex. Metal loss is often considered to occur by physical deformation of the surface. Shearing or fracture may occur subsequently as deformed areas are impacted by the erodent.

Factors controlling the rate of metal loss are related to the quantity, impact angle, speed, and density of the erodent. Metal loss is a strong function of erodent kinetic energy. Erosion damage may be roughly considered inversely proportional to alloy hardness. Usually, in the absence of significant corrosion, the harder the alloy, the more resistant it will be to attack. Erodent velocity has a very important effect on metal loss. Doubling velocity may increase metal loss by a factor of 4 or more. If particulate matter is entrained, mass, size, and particle-size distribution all affect attack.

Critical factors

Feedwater system.   Impingement of high-speed, turbulent water causes most attack. Entrained particulate matter may accelerate metal loss. If gases become mixed with fluids, cavitation (see Chap. 18, "Cavitation") is more likely than "simple" erosion.

Turbine.   Turbines are subject to erosion caused by impingement of solid particles and condensed water droplets. Solid-particle erosion is caused by exfoliated oxides, primarily from superheaters, reheaters, and main steam piping, impacting turbine components. Oxides are often dislodged by stresses associated with thermal transients. Impingement of condensed steam droplets produces wastage on latter-stage buckets, drain lines, and piping. If condensed fluids have low pH, erosion-corrosion becomes more significant.

Identification

Preboiler.   Erosion caused by turbulent water flow in pumps usually produces smooth grooves, craters, and general thinning (Fig. 17.9). The oxide layer in eroded regions is thinner or absent, usually is a different color, and has a smoother surface texture than in unattacked adjacent regions. A drop of acid placed on freshly eroded carbon steel will reveal

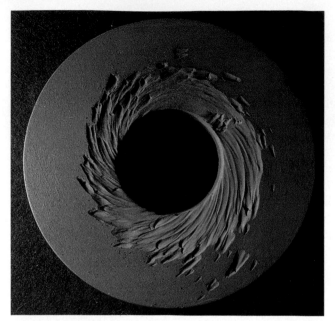

Figure 17.9 Feedwater-pump spacer eroded by turbulent high-speed water. Flow patterns are obvious.

bright metal in a matter of seconds. Unattacked surfaces usually require considerable exposure to reveal bare metal. Flow patterns are often obvious.

Unless particulate matter is entrained, close visual inspection will reveal no evidence of individual scratches, sleeks, or dents. Usually, such damage is apparent only when a larger amount of hard particulate matter is entrained within the carrying liquid.

Afterboiler.   Erosion caused by impingement of water droplets in latter turbine stages produces general wastage, which is most noticeable on leading edges of buckets (Fig. 17.10). Blade edges are marked by fine, transverse serrations and grooves (Fig. 7.11). Cone-shaped projections may rise from surfaces (Fig. 17.12). Often these cones will be tilted so that conical axes parallel impingement direction. Hence, cones are usually present on leading bucket faces and are less numerous or absent on back blade faces. Similar impingement cones can occur on mild steel surfaces such as turbine drain lines (Fig. 17.13).

Damage caused by hard-particle erosion causes tearing and microdenting of leading bucket edges. Striations and tilted cones are usually absent. A ragged feather edge usually develops.

Figure 17.10 Ragged leading edges of final-stage turbine blades. Damage was caused by water-droplet impingement.

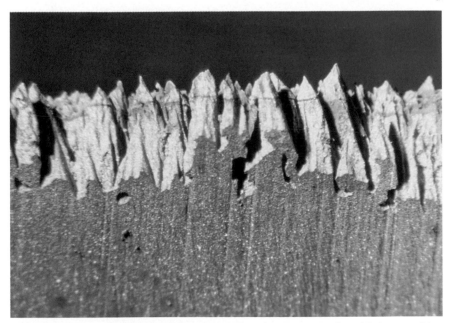

Figure 17.11 Fine, striated grooves on front face of final-stage turbine blade caused by water-droplet impingement. (Magnification: 7.5×.)

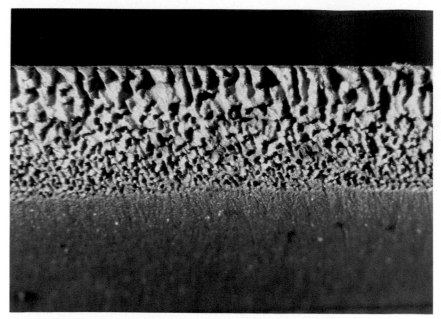

Figure 17.12 Dimpling of back edge of bucket surface shown in Fig. 17.11.

Elimination

Lessening attack requires eliminating the erodent, decreasing erodent kinetic energy, shielding surfaces, or substituting alloys that are more erosion-resistant. If pumps, valves, and piping are sized, designed, and operated properly, attack is rare. Turbine erosion problems can be more troublesome.

Figure 17.13 Erosion on mild steel turbine drain line. Surface is festooned with small, conical projections pointing toward the direction of impingement.

Magnetite exfoliation in superheaters, reheaters, and steam-transfer lines (see Chap. 1, "Water-Formed and Steam-Formed Deposits") is the major cause of solid-particle erosion in turbines. The exfoliation process is lessened when thermal stresses and tube temperatures are reduced. If superheater and reheater tubing is old, it is likely that tubes will contain considerable thermally formed oxide. Tube replacement or chemical cleaning may be necessary to remove thick oxide layers. Turbine screens and shielding devices should be maintained in good repair.

Alloys used in fabricating turbine buckets are quite similar worldwide, consisting of chromium stainless steels. However, cobalt-based alloys, titanium, and proprietary metals are gaining wider usage. Cobalt-based erosion-shield alloys have been used in latter stages.

Corrosion accelerates erosive metal loss. Almost all metals contacting steam or water experience some corrosion. Thus all erosion in boiler systems containing water or steam is an erosion-corrosion process. It is only when corrosion has a relatively small effect on metal loss that the process can be called "erosion." If erosion predominates, chemical inhibition can usually do little to reduce attack. But if corrosion is significant, the judicious use of inhibitors and/or pH modification may be beneficial.

Cautions

Cavitation is closely related to erosion. Damage occurs where high-velocity turbulent fluids are present. Cavitation damage may superficially resemble hard-particle erosion. However, cavitation does not occur in steam and is not likely to produce smooth, undulating, or grooved surfaces.

Attack by concentrated chelant increases substantially with flow velocity. Grooving and general thinning from chelant corrosion strongly resemble erosion alone. Chelant attack in feedwater lines, steam drums, and generating tubes is frequently confused with erosion.

Related problems

See also Chap. 5, "Chelant Corrosion"; Chap. 18, "Cavitation"; and the section titled Falling-Slag Erosion in this chapter.

Water-Side and Steam-Side Erosion

Erosion is very rare in water- and steam-cooled tubes. However, wastage will occur where flow is restricted by foreign objects, collections of scale, and the like. Usually, however, other associated problems such as overheating will cause failure before erosion can cause severe damage (see Case History 17.4).

CASE HISTORY 17.1

Industry:	Utility
Specimen Location:	Superheater
Specimen Orientation:	Vertical
Years in Service:	5
Water-Treatment Program:	Coordinated phosphate
Drum Pressure:	2400 psi (16.5 MPa)
Tube Specifications:	2½ in. (6.4 cm) outer diameter, SA-213
Fuel:	Pulverized coal

A series of failures of superheater tubes occurred almost 2 years after extensive retubing. Nearby tubes contained forming defects present as deep fissures. These fissures opened in service, causing steam cutting of adjacent tubes (Fig. 17.5).

It is remarkable that failures similar to this one occurred almost 2 years earlier, and for essentially the same reason. In spite of the extensive retubing, some defective tubes were missed. Tubes containing deep fissures remained in service for at least 2 years before failure occurred. When these defective tubes finally failed, extensive steam cutting of nearby tubes resulted.

CASE HISTORY 17.2

Industry:	Utility
Specimen Location:	Primary-superheater pendant
Specimen Orientation:	Vertical pendant
Years in Service:	25
Water-Treatment Program:	Coordinated phosphate
Drum Pressure:	1800 psi (12.4 MPa)
Tube Specifications:	2⅜ in. (6.0 cm) outer diameter
Fuel:	Coal

A superheater tube failed at a pendant U-bend. Failure was caused by corrosion-fatigue cracking. Leakage was relatively slight before opposite legs were steam-cut (Fig. 17.6). More tubes had to be replaced because of chain-reaction failures associated with steam cutting than because of the precipitating corrosion fatigue.

CASE HISTORY **17.3**

Industry:	Refining
Specimen Location:	Flue-gas cooler used to preheat boiler feedwater
Specimen Orientation:	Vertical
Years in Service:	4
Water-Treatment Program:	Polymer
Drum Pressure:	150 psi (1.0 MPa) in cooler tubes
Tube Specifications:	2 in. (5.1 cm) outer diameter

A longitudinally split section of flue-gas cooler used to preheat boiler feedwater contained helical grooves on internal surfaces (Fig. 17.14). Some grooves penetrated the tube wall, producing helical cracks, while other grooves went only one-third of the way through the wall. Grooves were undercut in the direction of gas flow.

Erosion occurred along a helical coil inside the tube. The helix was used to increase turbulence and eliminate deposits. High-velocity flue-gas flow became sufficiently turbulent at the helix to cause erosion failures.

Figure 17.14 Spiral grooves cut into internal surface of flue-gas cooler by erosive gas flow. A spiral, stainless coil present inside the tube caused increased localized turbulence.

CASE HISTORY 17.4

Industry:	Pulp and paper
Specimen Location:	Superheater U-bend
Specimen Orientation:	Vertical
Years in Service:	10 months
Water-Treatment Program:	Phosphate
Drum Pressure:	570 psi (3.9 MPa)
Tube Specifications:	1½ in. (3.8 cm) outer diameter
Fuel:	Black liquor

Severe localized metal loss from internal surfaces caused a perforation in a U-bend 10 months after tube installation. This was the only tube that was affected in this manner. An elliptical pattern of metal loss was present in the immediate vicinity of the perforation (Fig. 17.15). Islands of intact tube wall were present in the wasted region.

Failure was caused by impingement of high-velocity fluid on tube surfaces. It is likely that an object lodged in the tube on the upstream side of the bend constrained the water flow into a narrow channel. Erosion was induced locally by increased water speed and greater turbulence near the foreign object.

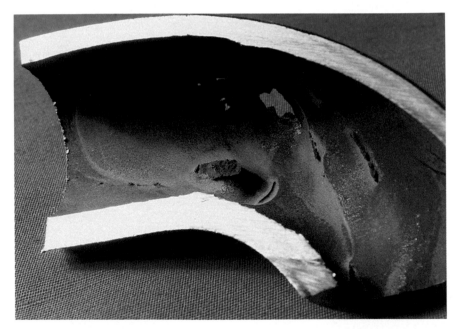

Figure 17.15 Elliptical metal-loss pattern and intact island of tube metal at perforated superheater U-bend. Erosion was caused by an object left in the tube 10 months earlier.

CASE HISTORY **17.5**

Industry:	Utility
Specimen Location:	Corner tube in a waterwall
Specimen Orientation:	Vertical
Years in Service:	19
Water-Treatment Program:	Coordinated phosphate
Drum Pressure:	2250 psi (15.5 MPa)
Tube Specifications:	3 in. (7.6 cm) outer diameter
Fuel:	Pulverized coal

A rupture occurred at a repair weld. Welding was used to increase lost wall thickness to about 50% of the specified wall thickness. These welds built up metal loss from fire-side surfaces.

External-surface metal loss was caused by erosion due to the impingement of fly ash entrained in flue gas. Deep grooves were cut in the tube wall at membrane gaps (Fig. 17.8).

This boiler had a history of slagging problems. High solids and silica concentration in the fuel accelerated attack.

CASE HISTORY **17.6**

Industry:	Pulp and paper
Specimen Location:	High-pressure section of condensing turbine, last row, condensing section
Specimen Orientation:	Horizontal turbine shaft
Years in Service:	6
Water-Treatment Program:	Chelant
Drum Pressure:	45-MW turbine, 3600 rpm, 830°F (450°C) superheated steam

Turbine buckets were damaged severely on leading edges. Edges were striated and grooved (Fig. 17.11). Erosion was caused by high-velocity fluids in which water droplets were entrained. Erosion damage due to water-droplet impingement is not uncommon in late stages of condensing turbines.

A 65-psi (0.45-MPa) steam-extraction, nonreturn valve failed to seat properly during an electrical turbine trip. The 65-psi (0.45-MPa) header emptied through the turbine (vacuum condition). The turbine speed increased to 5000 rpm (design 3600 rpm) before manual shutdown.

CASE HISTORY 17.7

Industry:	Utility
Specimen Location:	Division wall in a circulating fluidized-bed boiler
Specimen Orientation:	Vertical
Years in Service:	18 months, but only operating about 6 months because of numerous maintenance shutdowns
Water-Treatment Program:	Chelant
Drum Pressure:	1275 psi (8.8 MPa), ~1700°F (930°C) external-surface temperature
Tube Specifications:	3 in. (7.6 cm) outer diameter, mild steel
Fuel:	Furnace bed of sand, lime, and wood chips

The section has a massive longitudinal rupture in a zone of severe metal loss from external surfaces (Fig. 17.16). Failure was caused by external-surface erosion. Microstructural examination revealed that hard particulate matter (sand) was entrained in bed gases. Impingement of sand against the tube reduced wall thickness, resulting in severe thinning. Rupture occurred when internal pressure exceeded the yield strength of the thinned tube.

The forced-draft fan maintained a 60- to 70-in. (15- to 17.5-cm) pressure head of water to fluidize the bed. No refractory or other surface protective devices such as studs or wear bars were present at the failure site.

Figure 17.16 Ruptured division-wall tube from fluidized-bed boiler. Attack was caused by sand abrasion in the bed.

CASE HISTORY 17.8

Industry:	Refining
Specimen Location:	Condensate return header
Specimen Orientation:	Elbow (horizontal to vertical)
Years in Service:	Unknown (more than 5)
Tube Specifications:	6½ in. (16.5 cm) outer diameter, mild steel

Internal surfaces were covered in some areas with a smooth, black magnetite layer. Severe metal loss was present in large, distinct patches (Fig. 17.17). The metal loss produced smooth, mildly rolling contours free of deposits or corrosion products. Flow-oriented striations were present at the perimeter of the affected area (Fig. 17.18).

Condensate lines frequently suffer corrosion by carbonic acid. However, this tube shows evidence of erosion only. Metal loss is localized and distinctly flow-oriented. Such patterns are characteristic of erosion. Evidence suggests that magnetite particles entrained in the condensate contributed to metal loss.

Figure 17.17 Localized metal loss in a bend of a condensate return header.

Figure 17.18 Erosion patterns at edge of wasted area. As in Fig. 17.17. Note flow patterns. (Magnification: 7.5X.)

CASE HISTORY 17.9

Industry:	Pulp and paper
Specimen Location:	Economizer
Specimen Orientation:	Curved, predominantly vertical
Years in Service:	8
Water-Treatment Program:	Phosphate
Drum Pressure:	1200 psi (8.3 MPa)
Tube Specifications:	2 in. (5.1 cm) outer diameter, mild steel
Fuel:	Black liquor

Substantial thinning of the external surface of the tube wall is present at the inner curvature of a bend in an economizer tube. A ragged rupture perforates the wall in a zone of severe thinning (Fig. 17.19). Away from the rupture, surfaces are relatively unattacked.

The rupture was caused by erosive wall thinning. Normal internal pressures could no longer be contained at the eroded site and the tube ruptured.

Erosion was caused by impingement of hard particulate matter entrained in flue gases. There was evidence of mild cold-end corrosion on all external surfaces.

Figure 17.19 Rupture in an econo-mizer tube caused by severe local-ized external-surface erosion as-sociated with particulate matter entrained in flue gas.

CASE HISTORY **17.10**

Industry:	Utility
Specimen Location:	Front convection wall
Specimen Orientation:	Vertical
Years in Service:	8
Drum Pressure:	2500 psi (17.2 MPa)
Tube Specification:	2¾ in. (7.0 cm) outer diameter, SA201 A1
Fuel:	Coal

A tube failed at a weld support. Escaping water cut away the support and perforated the adjacent tube (Fig. 17.20).

The question that naturally arises in multiple failures is which failure occurred first. If erosion is severe, the original failure can be entirely eradicated. Luckily, other nearby supports were included with the received section. Careful visual inspection revealed small corrosion-fatigue cracks at poorly fused support welds. Upon sectioning and microscopic examination, small fissures were located near the original failure; these fissures were almost identical to (but deeper than) those found at adjacent braces. Hence, the original failure was likely caused by corrosion fatigue at a weld support.

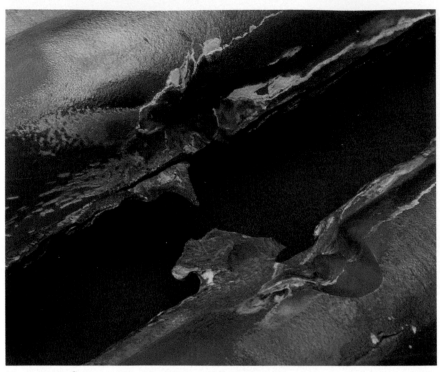

Figure 17.20 Steam cutting of utility convection tubes. The original failure occurred at a welded support bracket and was due to corrosion fatigue. The adjacent tube was breached by escaping steam from the original failure.

Cavitation

Locations

Cavitation is favored anywhere low-pressure regions form in water. Abrupt pressure changes and turbulent flow promote attack. Damage may occur only where water contacts surfaces. Pump impellers, and other pump components, are the most commonly attacked boiler component. Blowdown lines and valves are also frequently affected. Pump impellers are usually damaged on the suction side, and valves show wastage on discharge sides. Less commonly, condensate lines and turbine components are attacked.

General Description

Cavitation is a process whereby small vapor spaces rapidly form and collapse in a fluid. Pressure differences in the liquid cause vapor-bubble formation. The liquid actually boils at the reduced pressure. The steam bubbles quickly collapse, producing microjets that impinge on metal surfaces. The damage may affect only the normally protective oxide layer, or in severe cases, may directly attack the underlying metal, physically dislodging less resistant alloy phases.

Energy is required to form a cavitation bubble. Part of that energy is consumed in creating the bubble surface. Since it takes less energy to form the bubble on a preexisting surface, cavitation bubbles form most readily on existing surfaces. Pressure may be lowest and turbulence highest at or near moving surfaces. Further, surface discontinuities afford easy bubble-nucleation sites. Bubble formation and collapse may occur in just a small fraction of a second. Each bubble collapse produces a relatively small amount of damage. Damage accumulates during thousands of cycles. Once surface irregularities are formed, attack will tend to concentrate at damage sites, eventually producing deep, localized attack.

Critical Factors

Pump cavitation is often caused by too high a pressure differential between suction and discharge sides. Insufficient head pressure is usually the precipitating cause. Throttling on the suction side of pumps promotes a large pressure differential. Gas entrainment due to leaking packing, decomposition of water chemicals, and gas effervescence can also promote bubble

Figure 18.1 Longitudinally split blowdown line with severe localized metal loss caused by cavitation.

formation. In a surprising number of cases, incorrectly sized impellers and other pump components cause difficulties.

Blowdown lines are especially susceptible to damage if flow is excessive and the discharge direction abruptly changes. Attack usually occurs during intermittent manual blowdown when flow direction is severely changed in pipe tees and elbows. Attack can be intense if blowdown rate is high and lines are undersized (Fig. 18.1).

Alloy composition also influences attack. Soft, ductile metals and brittle, low-strength alloys such as gray cast iron are easily attacked (Fig. 18.2). Alloys such as chromium stainless steels are resistant to attack in many environments.

Turbulent flow and abrupt pressure changes promote attack. In many cases, cavitation is a threshold phenomenon. Cavitation does not simply gradually decrease in intensity, but ceases altogether below some critical turbulence level.

Identification

Cavitation damage produces localized areas of jagged metal loss. Undercutting is usually pronounced. Close visual inspection reveals surfaces having a spongy, honeycombed texture. Attacked regions may or may not

Figure 18.2 Pump-housing section severely attacked by cavitation. Material of the pump-housing section is gray cast iron.

Figure 18.3 A cast steel feedwater-pump impeller severely damaged by cavitation. Note how damage is confined to the outer edges of the impeller where vane speed was maximum.

be covered with corrosion products. However, if attack is active shortly before the section is inspected, corrosion products will be minimal. Although corrosion usually accelerates metal loss, attack can occur in its absence. Even glass can be attacked by cavitation.

A striking feature of cavitation damage is the localized nature of attack. Wastage is most severe on pump impeller vanes near the outer periphery where impeller speed, and presumably turbulence, is highest (Fig. 18.3). Attack is most severe on trailing vane edges, where low-pressure areas are formed (Fig. 18.4). Pump shafts, for example, can be attacked locally where surfaces are exposed to turbulent conditions (Fig. 18.5). Adjacent surfaces are usually only lightly polished or apparently totally free of damage. Edges, corners, and projections all intensify turbulence, provide bubble-nucleation sites, and become preferred damage regions.

If components are vibrating, damage may be present on all surfaces in contact with water. However, vibratory modes involving movement in a single plane are more common and produce attack on opposite sides of the component.

Cavitating pumps can sometimes be recognized by the sounds they

Figure 18.4 Vane damage on low-pressure sides of small, special-purpose, bronze feedwater-pump impeller. Some vanes have been penetrated.

Figure 18.5 Highly localized cavitation damage on steel feedwater-pump shaft. Adjacent regions are free of metal loss.

make. Active cavitation can sound like the impact of stones against metal surfaces. However, pump noise and vibration usually mask these sounds.

Elimination

Cavitation damage can be reduced by design, alloying, coating, and/or surface finishing. Design strategies concentrate on reducing turbulence, vibration, and rapid pressure changes. Maintaining sufficient head pressure, preventing packing leaks, and discharge-side throttling all reduce pump damage. Smoothing of impeller surfaces and use of elastomeric coatings prevent damage by reducing nucleation sites and by absorbing the implosion energy of bubbles, respectively.

Hard, resistant alloys including 18-8 stainless steels are often recommended. Plasma-coating and surface-hardening techniques have had limited success in reducing attack.

Cautions

Cavitation damage resembles acid corrosion. Although somewhat similar to other impingement and erosion phenomena, cavitation produces distinctly different wastage. Because of the unique characteristics of this damage, microscopic observations often conclusively show that cavitation has occurred.

Related Problems

See also Chap. 7, "Low-pH Corrosion during Acid Cleaning," and Chap. 17, "Erosion."

CASE HISTORY 18.1

Industry:	Chemical manufacturing
Specimen Location:	Blowdown line threaded into a tee
Specimen Orientation:	Horizontal
Years in Service:	6
Water-Treatment Program:	Phosphate
Drum Pressure:	1200 psi (8.3 MPa)
Tube Specifications:	2½ in. (6.4 cm) outer diameter, mild steel
Fuel:	No. 6 fuel oil, natural gas

Severe, localized wastage on internal surfaces caused a blowdown line to fail. A perforation occurred near the attachment of the pipe to a tee (Fig. 18.1). Attack was confined to spongy, jagged patches of loss; surrounding surfaces were free of significant deterioration.

Failure occurred during manual blowdown. Blowdown had been increased due to feedwater contamination from the preboiler system.

CASE HISTORY 18.2

Industry:	Pulp and paper
Specimen Location:	Boiler feedwater-pump impeller
Specimen Orientation:	Vertical
Years in Service:	1
Water-Treatment Program:	Phosphate
Drum Pressure:	1200 psi (8.3 MPa)
Impeller Specifications:	10½ in. (27 cm) diameter, 7 vanes, cast steel
Fuel:	Black liquor

Feedwater supply could not meet boiler demand. The feedwater impeller was severely wasted, and the vanes were almost gone. Attack was confined to the periphery of the impeller (Fig. 18.3).

Wastage was apparently caused by cavitation induced by insufficient head pressure.

CASE HISTORY 18.3

Industry:	Steel
Specimen Location:	Condensate line
Specimen Orientation:	Horizontal
Years in Service:	3
Water-Treatment Program:	Polymer
Drum Pressure:	900 psi (6.2 MPa)
Tube Specifications:	4½ in. (11.4 cm) outer diameter
Fuel:	Natural gas

A condensate line developed several small perforations. When opened, the line was riddled with many irregular pits (Fig. 18.6). Some pits mutually intersected, forming more general areas of metal loss. Close visual inspection of pits revealed jagged, irregular internal-surface contours.

Pits strongly resembled oxygen corrosion (See Chap. 8, "Oxygen Corrosion"). However, microscopic examination showed evidence of severe plastic deformation and mechanical shock within, and only within, pits. Further investigation revealed a history of steam hammer and other mechanical vibration of this line.

It is possible that oxygen pits were present before cavitation became severe. However, it is clear that cavitation played a role in deepening these pits and causing final perforation.

Figure 18.6 Heavily pitted condensate line. Attack resembles oxygen pitting, but was caused (at least in part) by cavitation.

Forming Defects

Locations

Forming defects that affect serviceability are generally found in tubes. But beyond this simple fact, nothing can be said about locations, as defective material can be inadvertently placed in any tubed region.

General Description

Failures resulting from manufacturing defects are relatively rare. In fact, manufacturing defects account for less than 1% of total failures examined. Of those that do occur, the two types of defects that are fairly common include seam or lap defects in seamless tubing, and deficient welds in welded tubing.

Seam defects are crevices in steel that are closed but not metallurgically bonded. They may occur in unwelded tubes as a consequence of the presence of internal voids (pipe) or cracks in the ingot from which the tube was formed. Seam defects can also be caused by faulty methods of steel rolling in the steel mill.

Deficiencies in welded tubes are often caused by incomplete fusion of the weldment during the manufacturing process.

Critical Factors

The critical factors resulting in the use of construction materials having significant forming defects include insufficient adherence on the part of the manufacturer to specified fabrication practices or quality control practices and, possibly, insufficient adherence on the part of the boiler manufacturer to specified quality control practices.

Identification

Seam or lap defects appear in a transverse cross section as straight or gently curving cracks, which may run longitudinally for some distance along the tube wall (Fig. 19.1). Although the defect commonly originates at a surface, it may be difficult to see since the surfaces of the defect are commonly covered with a layer of iron oxide or nonmetallic inclusions.

Deficient welds in welded tubing can often be identified visually once coverings of corrosion products or deposits are removed. The deficiency will be apparent as an intermittent or continuous opening or crevice that runs in a straight line down the tube wall.

Such defects in boiler tubes commonly result in stress rupture or fatigue-associated failures. In lower-temperature applications, these defects may induce pitting corrosion along the defect site.

Elimination

Generally, once construction is completed, there is no economical way of detecting specific manufacturing defects. If they exist, and are serious, their presence is generally revealed by failure of the defective component.

If failure of a component results from a manufacturing defect, it may be advisable to survey similar components for evidence of defectiveness. This usually requires the use of nondestructive testing techniques, especially ultrasonic testing.

Cautions

Thick-walled ruptures from overheating are sometimes incorrectly diagnosed in the field as material defects. Confirmation by metallographic examination may be required.

Related Problems

See also Chap. 2, "Long-Term Overheating," and Chap. 3, "Short-Term Overheating."

Figure 19.1   Curved defect originating on the internal surface of a seamless tube. This defect ran for 18 in. (45.7 cm) along the length of the tube. *(Courtesy of National Association of Corrosion Engineers.)*

CASE HISTORY 19.1

Industry:	Pulp and paper
Specimen Location:	Screen tube, recovery boiler
Specimen Orientation:	Vertical
Years in Service:	11
Water-Treatment Program:	Coordinated phosphate
Drum Pressure:	1200 psi (8.3 MPa)
Tube Specification:	2½ in. (6.3 cm) outer diameter
Fuel:	Black liquor

The longitudinal tube rupture apparent in Fig. 19.2 is 1½ in. (3.8 cm) long. Both visual and microstructural examinations reveal that the rupture occurred along a weld seam. Seam contours were clearly visible along the length of the tube.

Examination of cross sections taken at a distance from the through-wall rupture reveals that the unfused weld seam penetrates 40% of the tube-wall thickness, and is lined with dense iron oxide (Fig. 19.3). This is a weld-seam defect that occurred during tube manufacturing. An acid cleaning of the boiler caused the cavernous metal loss apparent near the surface in Fig. 19.3. External metal loss on the adjacent tube due to erosion from steam and water escaping from the through-wall rupture is apparent in Fig. 19.2.

Figure 19.2   Seam defect (top tube). External metal loss on the bottom tube is the result of erosion from mixed steam and water escaping through the rupture.

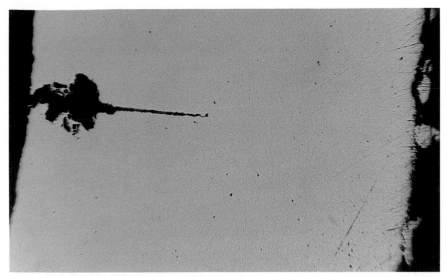

Figure 19.3 Cross-sectional view of weld-seam defect originating on the internal surface. Note enlargement of opening due to past acid cleaning of the boiler. (Magnification: 20×.)

CASE HISTORY 19.2

Industry:	Pulp and paper
Specimen Location:	Chill tube, power boiler, located in dead air space
Specimen Orientation:	Vertical
Years in Service:	25
Water-Treatment Program:	Phosphate
Drum Pressure:	900 psi (6.2 MPa)
Tube Specifications:	4 in. (10.2 cm) outer diameter

The massive failure illustrated in Fig. 19.4 occurred as the boiler was being brought on line. Drum pressure at the time of failure was 800 psi (5.5 MPa). An identical failure occurred a year earlier under the same conditions. The boiler is used intermittently and had been out of service for 5 months previous to this failure. The tube supplies a lower header in a waterwall.

Figure 19.5 illustrates the brittle rupture. Note the numerous parallel striations running along the fracture face.

Visual examinations of the internal surface revealed fissures along a welded seam aligned with the rupture. Microstructural examinations revealed that the rupture occurred along the weld seam. These examinations also disclosed deep fissures and pits along the seam resulting from exposure to acid during acid cleanings of the boiler.

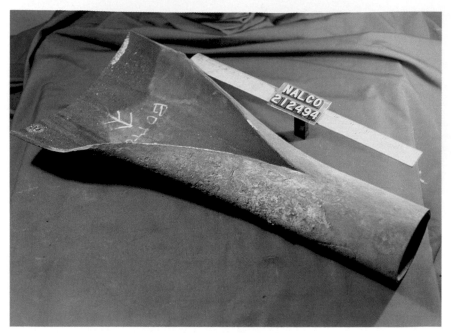

Figure 19.4 Massive rupture of tube along a weld-seam defect.

Figure 19.5 Edges of brittle rupture.

Apparently, the weld seam had not completely fused during tube fabrication. Acid collecting in the narrow crevice escaped neutralization. Consequently, when the boiler was restored to service following the acid cleaning, severe corrosion occurred along the seam, causing it to deepen. Eventually, stresses imposed by the cyclic operation of the boiler caused the crevice to grow by a corrosion-fatigue mechanism. When the crevice achieved a critical depth, fracture occurred through the remaining intact tube wall, which was incapable of supporting stresses imposed by internal pressure. Overheating had not occurred.

CASE HISTORY 19.3

Industry:	Utility
Specimen Location:	Secondary-superheater pendant
Specimen Orientation:	Vertical
Years in Service:	1
Water-Treatment Program:	Congruent control
Drum Pressure:	2500 psi (17.2 MPa)
Tube Specifications:	1¾ in. (4.4 cm) outer diameter, SA-213 T22, seamless tube
Fuel:	Coal

The failure illustrated in Fig. 19.6 is an 8-in.-long (20.3-cm-long), thick-walled fracture that originated on the internal surface (Fig. 19.1). Examination of the fracture face reveals a very smooth contour curving toward the external surface (Fig. 19.7). The extent of this smooth area is approximately 60 to 80% of the tube-wall thickness and runs 18 in. (45.7 cm) of the tube length. The fracture is covered with a very uniform layer of black iron oxide. The remaining 20 to 40% of the fracture surface adjacent to the external surface is irregular and lacks a distinct iron oxide covering.

Microstructural examinations revealed evidence of moderate overheating. Nonmetallic inclusions and the metal grains themselves are aligned along the contours of the smooth portion of the fracture.

This failure is a direct result of a manufacturing defect that extended for 60 to 80% of the original tube-wall thickness. Elevated stresses through the reduced tube-wall cross section, coupled with moderate overheating, caused final failure by stress rupture through the remaining intact tube wall. The defect appears to be an internal lap or seam, possibly associated with remnants of pipe (internal voids) in the steel ingot from which the tube was formed.

Rupture of the tube resulted in rapid, severe, localized wall thinning of adjacent tubes due to erosion from steam and water issuing from the rupture (see Case History 17.1). A second identical failure occurred in the superheater section 4 years later. It is remarkable that a tube with a defect of this magnitude could survive 5 years of service in a boiler of this pressure.

Figure 19.6 Failure along a massive manufacturing defect in the tube wall. *(Courtesy of National Association of Corrosion Engineers.)*

Figure 19.7 Smooth curved contour of defect. *(Courtesy of National Association of Corrosion Engineers.)*

CASE HISTORY 19.4

Industry:	Automotive
Specimen Location:	Steam line
Specimen Orientation:	Horizontal
Years in Service:	20
Water-Treatment Program:	Polymer
Drum Pressure:	210 psi (1.4 MPa), cyclic
Tube Specifications:	4½ in. (11.4 cm) outer diameter

The seam defect apparent in Fig. 19.8 required the replacement of a 25-ft (7.6-m) section of steam line. Ruptures in a similarly defective line had occurred a year earlier. A crack, which originated at the seam, penetrates the external surface for most of this 4-in. (10.2-cm) section.

Microstructural evaluations revealed decarburization of the metal, as well as the presence of iron nitride needles and nonmetallic inclusions in the weld-seam area. The through-wall crack followed the pathways of the nitride needles.

Figure 19.8 Seam defect in steam line.

The incompletely fused seam formed a narrow crevice that served as a stress-concentrating notch. Under the influence of cyclic pressure variations, corrosion-fatigue cracks propagated from the tip of the notch. The progress of these cracks was accelerated by the embrittling effect of the nitride needles present along the weld-seam zone. The combination of the diminished wall cross section resulting from an incompletely fused seam, the propagation of corrosion-fatigue cracks originating at the root of the open seam, and the presence of iron nitride needles resulted in a through-wall longitudinal failure of the line.

Welding Defects

Locations

A large, utility-size boiler may contain more than 50,000 welds. Each weld is a possible defect site. Largely as a result of the rigorous ASME code requirements for pressure vessels, weld failures account for only 2½% of boiler failures. Because of the relatively severe environments present in superheater and reheater sections, many of the weld failures that do occur happen in these regions.

General Description

The purpose of a weld is to join two metals by fusing them at their interfaces. A metallurgical bond is formed that provides smooth, uninterrupted, microstructural transition across the weldment. The weldment should be free of significant porosity and nonmetallic inclusions, form smoothly flowing surface contours with the section being joined, and be free of significant residual welding stresses.

A complete listing of possible weld defects is well beyond the scope of this manual. Rather, only the more common weld failures will be presented.

Critical Factors

All weld defects represent a departure from one or more of the features described above. However, it should be realized that welds are not perfect.

In consideration of this fact, all major welding codes allow for welding defects, but set limitations on the severity of the defect. An acceptable weld is not one that is defect-free, but one in which existing defects do not prevent satisfactory service.

Porosity

Identification.   *Porosity* refers to the entrapment of gas bubbles in the weld metal resulting either from decreased solubility of a gas as the molten weld metal cools, or from chemical reactions that occur within the weld metal (Fig. 20.1). Generally, porosity is not apparent from the surfaces of the weldment. The distribution of pores within the weldment can be classified as *uniformly scattered porosity, cluster porosity,* or *linear porosity.* Porosity near surfaces seems to have a significant effect on mechanical properties of the weld.

Elimination.   Porosity can be limited by using clean, dry materials, and by maintaining proper weld current and arc length.

Slag inclusions

Identification.   The term *slag inclusions* refers to nonmetallic solids trapped in the weld deposit, or between the weld metal and base metal. Slag forms by a high-temperature chemical reaction, which may occur during the welding process. These inclusions may be present as isolated particles, or as continuous or interrupted bands. Figure 20.2 illustrates a linear slag

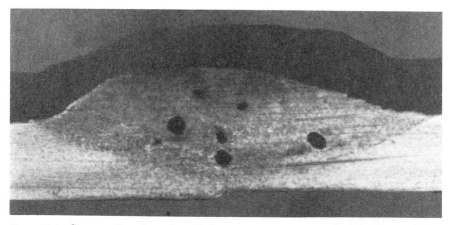

Figure 20.1 Cross section through weld showing severe porosity. This weld had been in low-pressure service for 40 years without failure. (Magnification: 2X.) *(Reprinted with permission from Helmut Thielsch,* Defects and Failures in Pressure Vessels and Piping, *New York, Van Nostrand Reinhold, 1965.)*

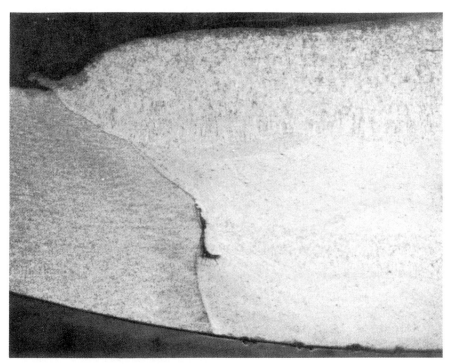

Figure 20.2 Cross section through a linear slag inclusion. This weld had been in 650-psi (4.5-MPa) service for 25 years. (Magnification: 5×.) *(Reprinted with permission from Helmut Thielsch,* Defects and Failures in Pressure Vessels and Piping, *New York, Van Nostrand Reinhold, 1965.)*

inclusion located at the weld-metal/base-metal interface. Slag inclusions are not visible unless they emerge at a surface.

Service failures are generally associated with surface-lying slag inclusions, or inclusions that are of such size that they significantly reduce cross-sectional area of the wall.

Elimination. The number and size of slag inclusions can be minimized by maintaining weld metal at low viscosity, preventing rapid solidification, and maintaining sufficiently high weld-metal temperatures.

Excess penetration (burnthrough)

Identification. The term *excess penetration* refers to disruption of the weld bead beyond the root of the weld. This disruption can exist as either excess metal on the back side of the weld, which may appear as "icicles" (Fig. 20.3), or concavity of the weld metal on the back side of the weld, sometimes referred to as "sink" (Fig. 20.4).

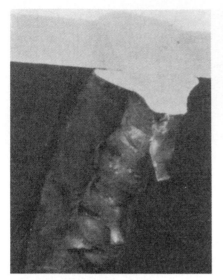

Figure 20.3 Burnthrough resulting in icicles on the underside of a butt weld. *(Reprinted with permission from Helmut Thielsch,* Defects and Failures in Pressure Vessels and Piping, *New York, Van Nostrand Reinhold, 1965.)*

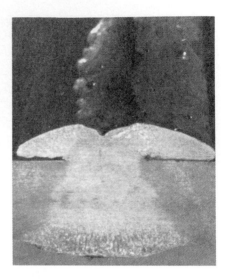

Figure 20.4 A burnthrough cavity formed on the backing ring of a tube weld. *(Reprinted with permission from Helmut Thielsch,* Defects and Failures in Pressure Vessels and Piping, *New York, Van Nostrand Reinhold, 1965.)*

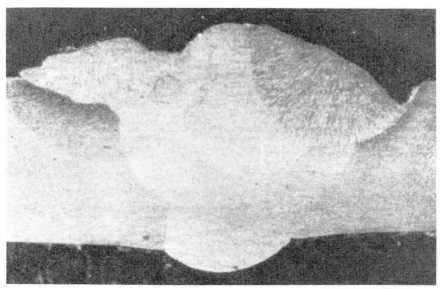

Figure 20.5 Cross section of welds containing severe undercut. (Magnification: 1.5×.) *(Reprinted with permission from Helmut Thielsch,* Defects and Failures in Pressure Vessels and Piping, *New York, Van Nostrand Reinhold, 1965.)*

The first of these is undesirable because the excess material can disrupt coolant flow, possibly causing localized corrosion downstream of the defect (water tube), or localized overheating downstream of the defect (steam tube).

The second of these, if severe, can cause root-pass cracking. Such cracks may not be revealed by radiographic examination. Concavity can also substantially reduce fatigue life, and excess concavity has been involved in thermal-fatigue failures.

Elimination.   Since disruptions of this type are frequently inaccessible for repair, elimination consists largely in preventing them from occurring in the first place. Excess penetration is frequently caused by improper welding techniques, poor joint preparation, and poor joint alignment.

Incomplete fusion

Identification.   *Incomplete fusion,* as the term implies, refers to lack of complete melting between adjacent portions of a weld joint. It can occur between individual weld beads, between the base metal and weld metal, or at any point in the welding groove.

Failures resulting from incomplete fusion of internal surfaces of the weld are infrequent, unless it is severe (amounting to 10% or more of the wall thickness). Incomplete fusion at surfaces, however, is more critical and can lead to failure by mechanical fatigue, thermal fatigue, and stress-corrosion cracking.

Elimination.   Incomplete fusion can occur because of failure to fuse the base metal or previously deposited weld metal. This can be eliminated by increasing weld current or reducing weld speed. Incomplete fusion can also be caused by failure to flux nonmetallic materials adhering to surfaces to be welded. This circumstance can be eliminated by removing foreign material from surfaces to be welded.

Related problems.   See the section titled Inadequate Joint Penetration in this chapter.

Undercut

Identification.   The term *undercut* refers to the creation of a continuous or intermittent groove melted into the base metal at either the surface (toe of the weld) (Fig. 20.5) or the root of the weld. Depending upon depth and sharpness, undercutting may cause failure by either mechanical or thermal fatigue.

Figure 20.6 Lack of penetration in the root of a butt weld in a boiler feedwater line. Sodium hydroxide concentrating in the resulting crevice led to caustic stress-corrosion cracking. Note fine cracks emanating from crevice. *(Reprinted with permission from Helmut Thielsch, De-fects and Failures in Pressure Vessels and Piping, New York, Van Nostrand Reinhold, 1965.)*

Elimination.   Serious undercutting may be repaired either by grinding or by depositing additional weld metal. Undercutting is generally caused by using excessive welding currents for a particular electrode or maintaining too long an arc.

Inadequate joint penetration

Identification.   Inadequate joint penetration, as the name implies, involves incomplete penetration of the weld through the thickness of the

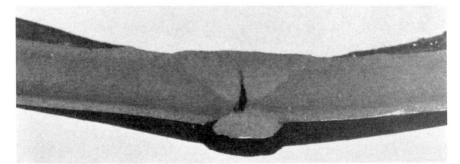

Figure 20.7 Cross section of circumferential butt weld. *(Reprinted with permission from Helmut Thielsch, Defects and Failures in Pressure Vessels and Piping, New York, Van Nostrand Reinhold, 1965.)*

joint (Fig. 20.6). It usually applies to the initial weld pass, or to passes made from one or both sides of the joint. On double-welded joints, the defect may occur within the wall thickness (Fig. 20.7).

Inadequate joint penetration is one of the most serious welding defects. It has caused failures in both pressure vessels and tube welds. Failures by mechanical fatigue, thermal fatigue, stress-corrosion cracking, and simple corrosion have been associated with this defect.

Elimination. Inadequate joint penetration is generally caused by unsatisfactory groove design, too large an electrode, excessive weld travel rate, or insufficient welding current.

Related problems. See the section titled Incomplete Fusion in this chapter.

Cracking

Identification. Cracks appear as linear openings at the metal surface. They can be wide, but frequently are tight. Such cracks are typically thick-walled and exhibit very little, if any, plastic deformation. Cracking of weld metal can be critical and has led to frequent service failures. Base-metal cracking (toe cracking, underbead cracking) is also critical and has caused service failures. Cracking may be either transverse or longitudinal (Fig. 20.8).

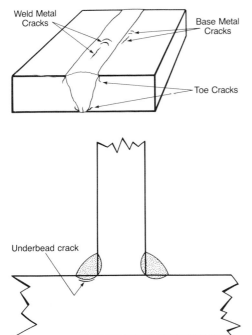

Figure 20.8 Typical crack locations in weldments.

Cracking in weldments can occur in several forms:

- **Hot cracking in weld deposit.** Forms immediately upon solidification of the weld metal.

- **Cold cracking in weld deposit.** Forms after the weld has cooled. Such cracks may form days after the welding procedure.

- **Base-metal hot cracking.** Forms upon solidification of the weld metal.

- **Base-metal cold cracking.** Forms after the weld metal has cooled.

Elimination. In general, cracking results from poor welding practice, inadequate joint preparation, use of improper electrodes, inadequate preheat, and an excessive cooling rate.

If cracking occurs in the weld metal, the following steps may prevent recurrence:

1. Decrease weld travel speed.
2. Preheat area to be welded, especially on thick sections.
3. Use low-hydrogen electrodes.
4. Use dry electrodes.
5. Sequence weld beads to accommodate shrinkage stresses.
6. Avoid conditions that cause rapid cooling.

If cracking occurs in the base metal, the following steps may prevent recurrence:

1. Preheat area to be welded, especially in thick sections.
2. Improve heat-input control.
3. Use correct electrodes.
4. Use dry electrodes.

Cautions. Final determination of the severity of a weld defect, and its influence on serviceability, requires the judgement of an experienced weld inspector.

Graphitization

Locations. Unalloyed steel and steel containing ½% molybdenum can be susceptible to graphitization. Such steels may be used in low-temperature sections of the superheater and reheater. Failures are more frequent in steam piping than in boiler tubing.

General description. The term *graphitization* refers to the spontaneous thermal conversion of the iron carbide constituent of steel to free carbon (graphite) and pure iron. This is a time/temperature phenomenon that typically occurs in a temperature range of 800 to 1200°F (427 to 649°C). The time required for graphitization to occur decreases with increasing metal temperature.

Failures due to graphitization in welded steels result from the formation of continuous chains of graphite that preferentially align along the low-temperature edge of the heat-affected zone in the base metal. These continuous chains or surfaces of graphite represent a brittle plane through which cracks may readily propagate. Unalloyed steels and steels containing ½% molybdenum can be affected. Failures by graphitization have caused substantial damage.

Critical factors. The critical factors leading to potential failure resulting from graphitization are the use of a susceptible material, welding of susceptible material, and exposure of the weldment to temperatures above the range of 800 to 1200°F (427 to 649°C) for a prolonged period.

A *susceptible material* can be defined as an unalloyed steel, or an alloyed steel containing ½% molybdenum that has been deoxidized in the steel-making process by aluminum in amounts exceeding 0.5 lb/ton (0.226 kg/T).

Identification. Failures due to graphitization may occur in carbon and carbon–molybdenum steels that have been welded and subsequently exposed to metal temperatures between 800 and 1200°F (427 and 649°C) for a prolonged period. Fractures occur in the heat-affected base metal approximately ¹⁄₁₆ in. (1.6 mm) from the weld interface. A nondestructive technique for confirming the occurrence of graphitization does not exist.

Elimination. Avoiding prolonged exposure of welded metal to the temperature range over which graphitization occurs is the best method of eliminating graphitization in suspect material. Specifying nonsusceptible material in the design phase of equipment has also been useful in eliminating failures by graphitization.

Additional considerations

Welding debris. Welding debris such as weld spatter, shavings, filings, chips from grinding tube ends, and even welding tools, have found their way into tubes as a consequence of tube repair. If this debris is not removed, it can cause partial blockage of coolant flow and result in overheating failures such as stress rupture. Such failures can occur months after the completion of the repair.

CASE HISTORY 20.1

Industry:	Utility
Specimen Location:	Superheater outlet header
Specimen Orientation:	Slanted
Years in Service:	7
Water-Treatment Program:	Ammonia, hydrazine
Drum Pressure:	3800 psi (26.2 MPa)
Tube Specifications:	2¼-in. (5.7 cm) outer diameter, SA-213 T22

The tube illustrated in Figs. 20.9 and 20.10 had been welded to the finishing-superheater outlet header in the penthouse. The specimen shows a brittle fracture at the weld, which apparently popped out of the header after the failure. Similar fractures had not occurred previously, but several additional cracked and leaking welds were discovered upon the inspection associated with this failure. The boiler was base-loaded and in continuous service except for yearly maintenance outages.

Visual examination of the weld and the fracture face revealed that large areas of the weld root were unfused. Radial striations on the fracture face originated at the unfused region.

The specimen had failed at the weld that joined the tube to the header. Cracks initiated at the weld root, where incomplete fusion had occurred, and propagated from this region by a corrosion-fatigue mechanism. The unfused portion of the weld formed sharp crevices, which acted as stress-concentration sites that locally elevated normal stresses.

Figure 20.9 Cross section of tube showing weld metal (top of tube).

Figure 20.10 Fracture surface showing tube wall (inner ring) and fractured weld metal (outer ring).

Cyclic stressing in this region of the boiler is frequently caused by differences in rates and directions of thermal expansion between the header and the tubes. Terminal tubes are often affected and can crack at the toe of the weld even in connections that have been properly welded. The thermal expansion and contraction stresses are often associated with start-up and shutdown.

CASE HISTORY 20.2

Industry:	Utility
Specimen Location:	Screen tube
Specimen Orientation:	45° slant
Years in Service:	20
Water-Treatment Program:	Coordinated phosphate
Drum Pressure:	1500 psi (10.3 MPa)
Tube Specification:	3 in. (7.6 cm) outer diameter
Fuel:	Coal

A "window" section had been cut out of the hot side of the tube illustrated in Fig. 20.11 and a new section welded into place from the external surface. Cross sections cut through the welded window revealed deep fissures associated with the welds (Fig. 20.12). The appearance of this defect suggests that entrapped slag, dirt, or flux prevented the formation of a sound metallurgical bond. Defects of this type can lead to through-wall fatigue cracking or corrosion fatigue due to the stress-concentrating effect of the fissure.

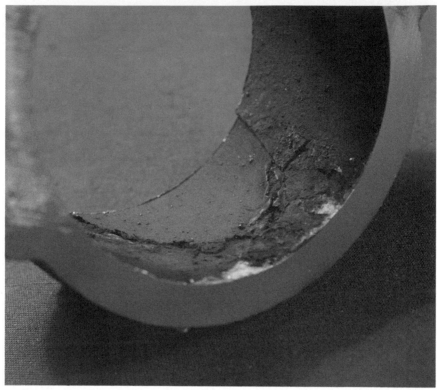

Figure 20.11   Internal surface of tube showing oval-shaped replacement window welded into place.

Figure 20.12 Cross section through window weld showing deep fissure in weld metal. (Magnification: 6×.)

CASE HISTORY 20.3

Industry:	Utility
Specimen Location:	Tube in area of cyclone burners
Specimen Orientation:	Vertical
Years in Service:	22
Water-Treatment Program:	Coordinated phosphate
Drum Pressure:	1500 psi (10.3 MPa)
Tube Specifications:	1½ in. (3.8 cm) outer diameter

The boiler from which the tube illustrated in Fig. 20.13 was taken had experienced recurrent overheating failures in the areas around the cyclone burners and the furnace floor. The boiler had been in peaking service for the last 6 years.

Figure 20.13 Profile of tube wall showing unfused fissure remaining after weld repair.

Failures were repaired by patch-weld overlays. The illustration indicates that the weld metal has not penetrated through the tube wall, leaving the thick-walled fissure caused by the in-service overheating unfused. These fissures operate as stress-concentration sites. Consequently, under the combined influence of normal service stresses and high operating temperatures, creep cracks (see Chap. 2, "Long-Term Overheating") formed at the tip of the fissure. Failure of the weld-repaired tube wall was imminent.

CASE HISTORY 20.4

Industry:	Utility
Specimen Location:	Screen wall
Specimen Orientation:	Vertical
Years in Service:	9
Water-Treatment Program:	Coordinated phosphate
Drum Pressure:	900 psi (6.2 MPa)
Tube Specifications:	2 in. (5.1 cm) outer diameter
Fuel:	Coal

Fig. 20.14 illustrates a massive fracture along the weld bead attaching the membrane to the tube wall. Microstructural examinations revealed an aligned chain of graphite nodules in the heat-affected zone immediately adjacent to the weld bead. The fracture had occurred through this chain. The proximity and alignment of the graphite nodules provided a weak plane where a fracture could readily propagate due to stresses imposed by internal pressure. This tube section had been exposed to metal temperatures in excess of 850°F (454°C) for a very long period of time.

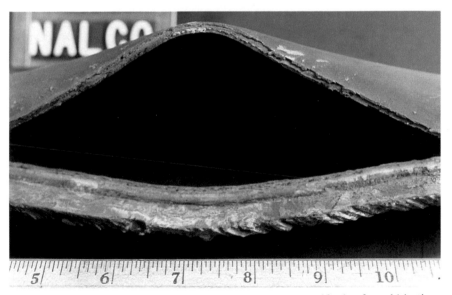

Figure 20.14 Massive fracture along weld bead resulting from weld-related graphitization. *(Courtesy of National Association of Corrosion Engineers.)*

Materials Deficiency

Locations

Materials deficiencies described in this chapter result from the inadvertent use of unalloyed steel in locations where alloyed steels are specified. Steam-cooled components, such as superheater and reheater tubes, are usually the areas of a boiler system that are affected.

General Description

With the possible exception of cast irons, the most commonly used industrial metal is unalloyed, plain carbon steel. Alloyed steels are used sparingly because of their expense. Most industrial and utility boilers use plain carbon steel in tubes that carry boiler water or a mixture of boiler water and steam. Low-alloy steels containing small amounts of chromium and molybdenum are commonly specified for steam-carrying tubes such as those in reheaters and superheaters because of the high temperatures at which these tubes operate. Chromium and molybdenum are added to retard the thermal degradation that occurs at these higher temperatures in unalloyed steel. Chromium is generally added in amounts of ¼ to 2¼% to retard rapid, high-temperature oxidation of the metal and to enhance creep resistance. Molybdenum is added in amounts of ¼ to 1% to increase creep resistance of the metal.

Critical Factors

The critical factors in materials deficiency are scaling temperature and susceptibility to creep. The *scaling temperature* is the temperature at which metals begin thermal oxidation at an accelerated rate (Fig. 21.1). Table 2.1 lists the steels commonly used in boiler tubes and their respective scaling temperatures.

Creep resistance is also a function of alloy content. Additions of chromium and molybdenum increase creep resistance.

Material deficiency occurs when a metal is used in an environment that exceeds the metal's thermal stability. Typically, this involves the use of unalloyed carbon steel in high-temperature superheater or reheater sections where normal operating temperatures would require the use of low-alloy steels.

Identification

Superheater and reheater tubes that exhibit excessive scaling and/or thick-walled ruptures (creep rupture) are suspect. Although low-alloy steels may exhibit these conditions if overheating has been severe, plain carbon steel will reach the same condition at lower metal temperatures. This makes plain carbon steel more susceptible to this type of failure in typical superheater and reheater environments. An examination of material specifications for the superheater or reheater sections will disclose the type of steel specified.

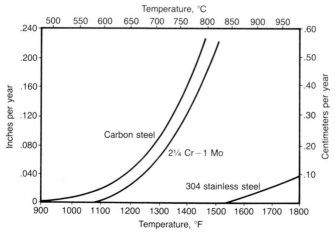

Figure 21.1 Approximate thermal oxidation (scaling) rates of carbon, low-alloy, and stainless steels in air.

It is difficult to identify positively tube metallurgy in the field. A section of the suspect tube should be submitted for alloy identification.

Elimination

The primary responsibility for prevention of misapplied material rests with the boiler manufacturer, boiler constructor, and purchaser of the boiler. Elimination requires close adherence to effective quality control procedures, beginning with equipment design and carrying through to equipment installation.

Cautions

Thermal-degradation failure (excessive scaling, creep rupture) of superheater and reheater tubes may occur both in properly specified and installed low-alloy steels, and in misapplied plain carbon steel tubes. Therefore, the occurrence of a thermal-degradation failure is not proof that a material has been incorrectly used in an application. If failures of this type occur in properly operating superheater or reheater sections, it is appropriate to suspect possible misapplication of materials.

Related Problems

See also Chap. 2, "Long-Term Overheating."

CASE HISTORY 21.1

Industry:	Pulp and paper
Specimen Location:	Secondary superheater, first pass
Specimen Orientation:	Vertical
Years in Service:	1½
Water-Treatment Program:	Coordinated phosphate
Drum Pressure:	1400 psi (9.7 MPa)
Tube Specifications:	2¾ in. (7.0 cm) outer diameter, carbon steel
Fuel:	Bark

The tube illustrated in Fig. 21.2 ruptured in service in a bark-fired boiler. Three similar failures had occurred over the previous year in this section of the superheater. The boiler is operated continuously except for semiannual maintenance outages.

Figure 21.2 Part of a thick-walled rupture in a superheater tube.

The rupture edges are thick, and have a chisellike contour. The external surface is covered with a barklike encrustation of thermally decomposed metal. The internal surface is smooth and free of deposits and corrosion products.

Microstructural examinations of the metal revealed the presence of graphite and spheroidized iron carbides in the tube wall at all locations. Intergranular oxidation, as well as thick layers of iron oxide (thermally decomposed metal), was also observed at all surfaces.

The tube had been operated above the scaling temperature [approximately 1000 to 1050°F (538 to 566°C) for plain carbon steel] for this metal over a long period. Previous failures, the first of which occurred a year after start-up of the boiler, indicate a possible imbalance between heat flux and coolant flow rate in this superheater section. However, failure is improbable had the tubes been fabricated from low-alloy steel (SA-213 T22).

CASE HISTORY 21.2

Industry:	Utility
Specimen Location:	Reheater
Specimen Orientation:	Vertical
Years in Service:	20
Water-Treatment Program:	Coordinated phosphate
Drum Pressure:	2875 psi (19.8 MPa)
Tube Specifications:	2¼ in. (5.7 cm) outer diameter, carbon steel
Fuel:	Coal

The thick-walled rupture shown in Fig. 21.3 is one of many similar failures that occurred sporadically in this section of the boiler. The boiler is presently base-loaded, but had been in peaking service previously.

The external surfaces are covered with thick layers of black iron oxide (thermally deteriorated metal) overlaid with light-colored deposits. The internal surface is also covered with thick layers of iron oxide.

Microstructural examinations of the metal revealed a dense population of graphite nodules and spheroidal iron carbides in the tube wall. Intergranular oxidation was apparent at all surfaces.

The failure of this tube is a direct result of operation in an environment that exceeded the thermal stability of the metal. A low-alloy steel tube (SA-213 T22) that had been butt-welded to the ruptured tube did not fail and had sustained only mild thermal deterioration.

Figure 21.3 Thick-walled rupture in a reheater tube. Note fissures on external surface aligned with the rupture.

Graphitic Corrosion

Locations

Feedwater pumps, water supply lines, valves, and other components made of cast irons (containing graphite) are attacked. Because cast irons are used mainly in preboiler regions, attack is found primarily in water-pretreatment and -transport equipment.

General Description

Graphitic corrosion is possible only in structures composed of cast irons containing graphite particles. Susceptible cast irons are nodular, malleable, and gray. Although frequently considered immune to graphitic corrosion, nodular cast iron and malleable iron are often corroded. Gray cast iron is more widely used and has more dramatic and recognizable corrosion characteristics than other cast irons.

A galvanic effect occurs between graphite particles embedded in the casting and the surrounding metal matrix when mildly aggressive water contacts surfaces. The graphite is cathodic to the adjacent metal. The metal portion of the casting corrodes. Eventually the casting is converted to rust containing graphite particles. Corrosion is accelerated if waters are mildly acidic, have high conductivity, are soft, and/or contain high concentrations of aggressive anion such as sulfate.

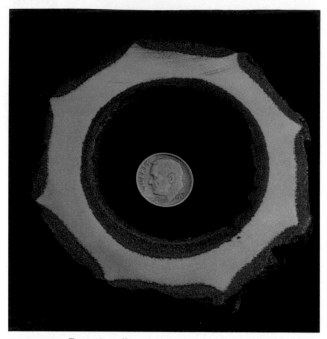

Figure 22.1 Pump impeller severely attacked by graphitic corrosion. Material of the impeller is gray cast iron. Note the gray areas on both internal and external surfaces where the metal has been converted to rust and graphite.

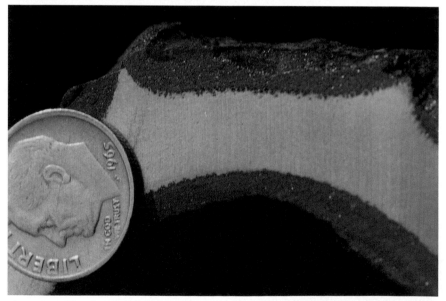

Figure 22.2 Note how uniformly metal is converted to corrosion product. Impeller vanes were lost in service because the graphitically corroded metal is brittle.

Critical Factors

Graphitic corrosion usually progresses slowly, taking many months or even years to produce significant attack. As pH decreases, attack quickens. Stagnant conditions promote attack, especially when waters contain high sulfate concentrations. Much attack occurs during idle periods. If turbulence is pronounced, (e.g., in pumps), corrosion products may be dislodged, also accelerating wastage.

Identification

Cast iron is converted to a soft mixture of iron oxides and graphite (Figs. 22.1 and 22.2). Pieces of corrosion product smudge hands and can be used to mark paper, just as if the corroded material were lead in a pencil (Fig. 22.3). Attack is often uniform, with all exposed surfaces corroded to roughly the same depth (Figs. 22.4 and 22.5). If localized deposits are present, especially those containing sulfate and chloride or other acidic species, corrosion may be confined to pockets. When attack is severe or prolonged, the entire component is converted to corrosion product. Surface contour and appearance are often preserved. Attack is usually not apparent until surfaces are probed or stressed.

Corroded areas can be broken with bare hands or by gentle taps with a hard implement. Craters can be dug in the soft, black corrosion product. Probing with a knife point can reveal the depth of penetration.

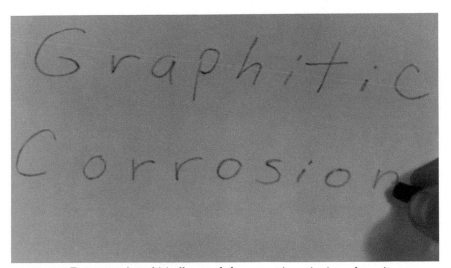

Figure 22.3 Fragment of graphitically corroded, gray cast iron pipe is used to write on paper, as if it were a lead pencil. The corrosion product is graphite flakes intermixed with rust.

Figure 22.4 Graphitically corroded valve butterfly. Original surface contours are preserved. Edges, which were completely converted to brittle corrosion product, have broken.

Figure 22.5 Severe graphitic corrosion of a gray cast iron feedwater-pump impeller. Patches of corrosion product have cracked and spalled, revealing uniform general wastage. Note the fracture in the impeller vane.

Elimination

Attack is reduced by alloy substitution, chemical treatment, and/or operational changes. Alloy substitution entirely eliminates graphitic corrosion if the corroded cast iron is replaced with alloys containing no graphite. The substitute alloy choice is dictated by requirements unique to each environment.

Raising water pH to neutral or slightly alkaline levels decreases attack, especially if relatively high concentrations of aggressive anions such as chloride and sulfate are present. The judicious use of chemical inhibitors may minimize deposits. When water flow is slight, such as during prolonged shutdowns and lengthy idle periods, attack increases. Stagnant conditions promote graphitic corrosion and should be avoided.

Cautions

Only cast irons containing graphite can corrode graphitically. Because of microstructural differences in graphite particle size and distribution, as well as other differences in alloy composition, attack is usually worst on gray cast irons.

Although pipes and other components may be severely corroded, they may not fail. The corrosion product has some mechanical strength, but is brittle. If corroded components are stressed, failure may occur catastrophically.

CASE HISTORY 22.1

Industry:	Food processing
Specimen Location:	Feedwater pump
Specimen Orientation:	Horizontal
Years in Service:	5
Water-Treatment Program:	Phosphate
Drum Pressure:	600 psi (4.1 MPa)
Tube Specifications:	8 in. (20.3 cm) diameter, 5 vanes, gray cast iron
Fuel:	Natural gas

A small feedwater-pump impeller was removed during a scheduled maintenance outage. The entire impeller surface was converted into soft, black corrosion products. In some areas, no trace of the original impeller alloy was left. Vane tips were totally converted to corrosion product. Vanes were cracked, and pieces of the corrosion product were dislodged, producing irregular contours (Fig. 22.5).

The impeller had been used intermittently, with greater than 50% idle time in the previous 2 years.

Dealloying

Locations

Dealloying, other than in cast irons (see Chap. 22, "Graphitic Corrosion"), usually occurs in copper-containing alloys. Corrosion is confined primarily to feedwater systems and afterboiler regions. Areas that can suffer attack include high-pressure feedwater heaters, bronze pump impellers, Monel steam strainers, and boiler peripherals such as brass pressure-gauge fittings. Condensers and heat exchangers are also frequently affected.

General Description

Dealloying is a corrosion process in which one or more alloy components are removed preferentially. The corroded region usually has a markedly different structure than the original alloy. However, macroscopic dimensions of the corroded part often remain unchanged. The process is also referred to as *selective leaching,* or *parting.* Obviously, dealloying can occur only in alloys containing two or more elements. Particularly susceptible alloys are cupronickels (in which nickel is removed) and brasses (in which zinc and aluminum are leached). The name given to a particular dealloying process derives from the leached element. For example, in common brasses where zinc is removed, dealloying is referred to as *dezincification.* In situations where nickel is removed from cupronickels, dealloying is referred to as *denickelification.*

Figure 23.1 Plug-type dezincification beneath a deposit in admiralty brass tube.

There are two commonly recognized forms of dezincification. Attack is either of a *plug* or *layer* type. Small localized areas of metal loss occur in the tube wall, producing plugs that can be blown out of pressurized tubes (Fig. 23.1). More general attack is called *layer-type dezincification* (Fig. 23.2). Cupronickels are more prone to layer-type attack than to plug-type deterioration. However, cupronickel wastage is usually slight compared to attack in brasses. It is likely that fundamental mechanisms of plug-type and layer-type attack are similar. However, plug-type attack can produce localized wastage rates of up to several hundred milliinches per year, while much milder attack is common with layer-type dealloying.

Destannification (loss of tin) can occur in gunmetal and some phosphor bronzes. This is especially the case in steam environments and in hot feedwaters. Monel steam strainers have been attacked when exposed to steam containing sulfur compounds at elevated temperatures (Fig. 23.3).

Figure 23.2 Layer-type dezincification of a brass pump component.

Dealuminification of duplex aluminum bronzes has been reported in waters having both high and low pH.

Critical Factors

Deposits, soft waters (especially those containing carbon dioxide), heat transfer, stagnant conditions, either high- or low-pH waters, and high-chloride waters accelerate most forms of dealloying in copper-containing alloys. Addition of small amounts of arsenic, antimony, or phosphorus to admiralty brasses has materially reduced the tendency to dezincify. However, attack may still occur under extreme conditions.

Identification

In cupriferous alloys, copper is almost never selectively removed. Rather, other elements are dissolved, leaving behind a comparatively soft, porous mass of copper. Attacked metal is relatively brittle and can be broken easily by impact or bending. Frequently, surfaces will be riddled with cracks, but will retain original dimensions. Attacked areas will usually

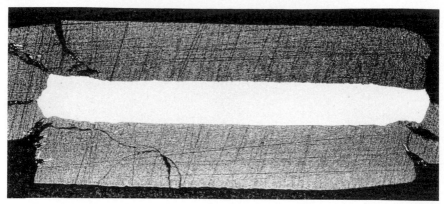

Figure 23.3 Cross section through a corroded Monel metal steam strainer. The dark area consists of oxides, sulfides, and elemental copper (Magnification: 15X.)

Figure 23.4 Peeling surfaces on a cupronickel high-pressure feedwater heater tube.

change to a deep red or salmon color of elemental copper (Figs. 23.1 and 23.2).

High-pressure feedwater tubes made of cupronickel may experience a unique form of dealloying. Air may gain access to shell-side surfaces during outages, causing considerable oxidation. Subsequent normal operation causes the surface oxides to be reduced to elemental copper. Eventually, sheets of oxide and/or reduced metal may peel off surfaces, giving the tubes an odd, sunburned appearance (Fig. 23.4). Alloys such as 70:30 cupronickels are particularly susceptible to this form of exfoliation. Alloys such as 80:20 are susceptible to a lesser extent. Such wastage is rare in 90:10 cupronickels.

Elimination

Surfaces should be kept free of deposits. In general, outages should be as short as is practical. Air contact should be prevented by steam or nitrogen blanketing. Water and steam quality must be controlled so that chloride- and sulfur-compound concentrations are minimized. Substitution of alternative alloys may be necessary. Use of inhibited grades of admiralty brass should be insisted upon when conditions dictate.

Related Problems

See Chap. 22, "Graphitic Corrosion."

CASE HISTORY 23.1

Industry:	Utility
Specimen Location:	Turbine steam strainer
Specimen Orientation:	Vertical
Years in Service:	16
Water-Treatment Program:	Phosphate
Drum Pressure:	1500 psi (10.3 MPa)
Specifications:	Monel wire, 0.15 in. (0.4 cm) diameter
Fuel:	Coal

A Monel turbine steam strainer was discovered to contain many broken elements. The metal was converted to oxide, sulfide, and elemental copper (Fig. 23.3). The deteriorated metal consisted of numerous elemental copper particles embedded in an oxide–sulfide matrix. Cracks in the wasted metal were lined with elemental copper.

Failure was attributed to carryover of sulfur-containing compounds in the steam. The system had a history of poor steam purity and carryover of boiler water into the superheater.

CASE HISTORY 23.2

Industry:	Utility
Specimen Location:	Inlet first pass of high-pressure feedwater heater
Specimen Orientation:	Horizontal
Years in Service:	7
Water-Treatment Program:	All volatile
Pressure:	400 psi (2.6 MPa)
Tube Specifications:	⅝ in. (1.6 cm) outer diameter, 70:30 cupronickel

Feedwater heater tubes were thinned by cyclic oxidation followed by reduction of oxides in service (Fig. 23.4). Wall thickness was reduced by as much as 15%.

During the previous 2 years, the boiler had experienced frequent outages in which air leaked into the heater shell and caused surface oxidation. Conversion of oxide to elemental copper occurred during normal operation.

Glossary

acid A compound that yields hydrogen ions (H^+) when dissolved in water.

alkaline 1. Having properties of an alkali. 2. Having a pH greater than 7.

alloy steel Steel containing specified quantities of alloying elements added to effect changes in mechanical or physical properties.

amphoteric Capable of reacting chemically either as an acid or a base. In reference to certain metals, signifies their propensity to corrode at both high and low pH.

anode In a corrosion cell, the area over which corrosion occurs and metal ions enter solution; oxidation is the principal reaction.

attemperator Apparatus for reducing and controlling the temperature of a superheated steam.

austenite A face-centered cubic solid solution of carbon or other elements in nonmagnetic iron.

austenitic stainless steel A nonmagnetic stainless steel possessing a microstructure of austenite. In addition to chromium, these steels commonly contain at least 8% nickel.

backing In welding, a material placed under or behind a joint to enhance the quality of the weld at the root. The backing may be a metal ring or strip; a pass of weld metal; or a nonmetal such as carbon, granular flux, or a protective gas.

base metal 1. In welding, the metal to be welded. 2. After welding, that part of the metal that was not melted.

black liquor The liquid material remaining from pulpwood cooking in the soda or sulfate papermaking process.

blowdown In connection with boilers, the process of discharging a significant portion of the aqueous solution in order to remove accumulated salts, deposits, and other impurities.

brittle fracture Separation of a solid accompanied by little or no macroscopic plastic deformation.

cathode In a corrosion cell, the area over which reduction is the principal reaction. It is usually an area that is not attacked.

caustic cracking A form of stress-corrosion cracking affecting carbon steels and austenitic stainless steels (300 series) when exposed to concentrated caustic — i.e., highly alkaline — solutions.

caustic embrittlement   An obsolete historical term denoting a form of stress-corrosion cracking most frequently encountered in carbon steels or iron–chromium–nickel alloys that are exposed to concentrated hydroxide solutions at temperatures of 200 to 250°C (400 to 480°F).

cavitation   The formation and instantaneous collapse of innumerable tiny voids or cavities within a liquid subjected to rapid and intense pressure changes.

cavitation damage   The degradation of a solid body resulting from its exposure to cavitation. This may include loss of material, surface deformation, or changes in properties or appearance.

cementite   A compound of iron and carbon, known chemically as iron carbide and having the approximate chemical formula Fe_3C.

chelating agent   An organic compound in which atoms form more than one coordinate bond with metals in solution.

cold work   Permanent deformation of a metal produced by an external force.

corrosion   The chemical or electrochemical reaction between a material, usually a metal, and its environment that produces a deterioration of the material and its properties.

corrosion fatigue   The process in which a metal fractures prematurely under conditions of simultaneous corrosion and repeated cyclic loading — fracture occurs at lower stress levels or fewer cycles than would be required in the absence of the corrosive environment.

corrosion product   Substance formed as a result of corrosion.

creep   Time-dependent deformation occurring under stress and high temperature.

creep rupture   See **stress rupture**.

dealloying   (see also **selective leaching**)   The selective corrosion of one or more components of a solid solution alloy. Also called *parting* or *selective leaching*.

denickelification   Corrosion in which nickel is selectively leached from nickel-containing alloys. Most commonly observed in copper–nickel alloys after extended service in fresh water.

dezincification   Corrosion in which zinc is selectively leached from zinc-containing alloys. Most commonly found in copper–zinc alloys containing less than 85% copper after extended service in water containing dissolved oxygen.

downcomer   Boiler tubes in which fluid flow is away from the steam drum.

ductile fracture   Fracture characterized by tearing of metal accompanied by appreciable gross plastic deformation and expenditure of considerable energy.

ductility   The ability of a material to deform plastically without fracturing.

economizer   A heat-exchange device for increasing feedwater temperature by recovery of heat from gases leaving the boiler.

erosion   Destruction of metals or other materials by the abrasive action of moving fluids, usually accelerated by the presence of solid particles or matter in suspension. When corrosion occurs simultaneously, the term *erosion-corrosion* is often used.

eutectic structure   The microstructure resulting from the freezing of liquid metal such that two or more distinct, solid phases are formed.

exfoliation   A type of corrosion that progresses approximately parallel to the outer surface of the metal, causing layers of the metal or its oxide to be elevated by the formation of corrosion products.

failure   A general term used to imply that a part in service (1) has become completely inoperable, (2) is still operable but is incapable of satisfactorily performing its intended function, or (3) has deteriorated seriously, to the point that it has become unreliable or unsafe for continued use.

fatigue   The phenomenon leading to fracture under repeated or fluctuating mechanical stresses having a maximum value less than the tensile strength of the material.

ferrite   Designation commonly assigned to alpha iron containing elements in solid solution.

ferritic stainless steel   A magnetic stainless steel possessing a microstructure of alpha ferrite. Its chromium content varies from 11.5 to 27%, but it contains no nickel.

fish-mouth rupture   A thin- or thick-lipped burst in a boiler tube that resembles the open mouth of a fish.

flux   (noun) 1. A substance, often a liquid, that is capable of dissolving metal oxides. 2. The rate of transfer of fluid, particles, or energy across a given surface.

gas porosity   Fine holes or pores within a metal that are caused by entrapped gas or by evolution of dissolved gas during solidification.

grain   An individual crystal in a polycrystalline metal or alloy.

grain boundary   A narrow zone in a metal corresponding to the transition from one crystallographic orientation to another, thus separating one grain from another.

graphitic corrosion   Corrosion of gray iron in which the iron matrix is selectively leached away, leaving a porous mass of graphite behind. Graphitic corrosion occurs in relatively mild aqueous solutions and on buried pipe fittings.

graphitization   A metallurgical term describing the formation of graphite in iron or steel, usually from decomposition of iron carbide at elevated temperatures. Not recommended as a term to describe graphitic corrosion.

gulping   Intermittent, brief passage of water from the steam drum of a boiler into the superheater caused by variable water levels.

heat-affected zone   In welding, that portion of the base metal that was not melted during welding, but whose microstructure and mechanical properties were altered by the heat.

hematite   A magnetic form of iron oxide, Fe_2O_3. Hematite is gray to bright red. The reddish forms are nonprotective, and their occurrence indicates the presence of high levels of oxygen.

hydrolysis   A chemical process of decomposition involving splitting of a bond and addition of the elements of water.

inclusions   Particles of foreign material in a metallic matrix. The particles are usually compounds (such as oxides, sulfides or silicates), but may be any substance that is foreign to (and essentially insoluble in) the matrix.

intergranular   Occurring between crystals or grains. Also, *intercrystalline*.

intergranular corrosion   Corrosion occurring preferentially at grain boundaries, usually with slight or negligible attack on the adjacent grains.

laning   The intentional or unintentional formation of a bypass or short circuit for furnace gases resulting in redistribution of heat-transfer rates.

lap   A surface imperfection having the appearance of a seam, and caused by hot metal, fins, or sharp corners being folded over and then being rolled or forged into the surface without being welded.

magnetite   A magnetic form of iron oxide, Fe_3O_4. Magnetite is dark gray to black, and forms a protective film on iron surfaces.

martensite   A supersaturated solid solution of carbon in iron characterized by a needlelike microstructure.

matrix   The principal phase in which another constituent is embedded.

microstructure   The structure of a metal as revealed by microscopic examination of the etched surface of a polished specimen.

mild steel   Carbon steel having a maximum carbon content of approximately 0.25%.

overheating   Heating of a metal or alloy to such a high temperature that its properties are impaired.

pearlite   A microstructural aggregate consisting of alternate lamellae of ferrite and cementite.

penetration   In welding, the distance from the original surface of the base metal to that point at which fusion ceased.

pH   The negative logarithm of the hydrogen ion activity; it denotes the degree of acidity or basicity of a solution. At $25°C$ ($77°F$), 7.0 is the neutral value. Decreasing values below 7.0 indicate increasing acidity; increasing values above 7.0 indicate increasing basicity.

pipe   The central cavity formed by contraction in metal, especially ingots, during solidification.

pitting   The formation of small, sharp cavities in a metal surface by corrosion.

plain carbon steel   Steel containing carbon up to about 2% and only residual quantities of other elements except those added for deoxidation. Also called *ordinary steel*.

residual stress Stresses that remain within a body as a result of plastic deformation.

riser Boiler tubes in which fluid flow is toward the steam drum.

root crack A crack in either a weld or the heat-affected zone at the root of a weld.

root of joint In welding, the portion of a weld joint where the members are closest to each other before welding. In cross section, this may be a point, a line, or an area.

root of weld The points at which the weld bead intersects the base-metal surfaces either nearest to or coincident with the root of joint.

scaling The formation at high temperatures of thick layers of corrosion product on a metal surface.

scaling temperature A temperature or range of temperatures at which the resistance of a metal to thermal corrosion breaks down.

seam On a metal surface, an unwelded fold or lap that appears as a crack, usually resulting from a discontinuity.

seam welding Making a longitudinal weld in sheet metal or tubing.

selective leaching Corrosion in which one element is preferentially removed from an alloy, leaving a residue (often porous) of the elements that are more resistant to the particular environment.

spalling The cracking and flaking of particles out of a surface.

stainless steel Any of several steels containing 12 to 30% chromium as the principal alloying element; they usually exhibit passivity in aqueous environments.

stoker-chain grate A device for conveying solid fuel across a furnace such that the grate acts as a burning platform.

stress Force per unit area, often thought of as force acting through a small area within a plane. Stress can be divided into components, normal and parallel to the plane, called *normal stress* and *shear stress*, respectively. *True stress* denotes the stress where force and area are measured at the same time. *Conventional stress*, as applied to tension and compression tests, is force divided by original area.

stress-corrosion cracking Failure by cracking under combined action of corrosion and stress, either external (applied) stress or internal (residual) stress. Cracking may be either intergranular or transgranular, depending on the metal and the corrosive medium.

stress raisers Changes in contour or discontinuities in structure that cause local increases in stress.

stress rupture (creep rupture) A fracture that results from creep.

synergism Cooperative action of discrete agencies such that the total effect is greater than the sum of the effects taken independently.

tensile strength In tensile testing, the ratio of maximum load to original cross-sectional area. Also called *ultimate strength*.

thermal fatigue The process leading to fracture under repeated or fluctuating thermally induced stresses having a maximum value less than the tensile strength of the material.

transgranular   Occurring through or across crystals or grains. Also, *intracrystalline* or *transcrystalline*.

tuberculation   The formation of localized corrosion products in the form of knob-like mounds called *tubercles*.

ultrasonic testing   A nondestructive test in which an ultrasonic beam is applied to sound-conductive materials having elastic properties; the test is used to locate inhomogeneities or structural discontinuities within the material.

underbead crack   A subsurface crack in the base metal near a weld.

undercut   In weldments, a groove melted into the base metal adjacent to the toe of a weld and left unfilled.

weld   A union made by welding.

weld bead   A deposit of filler metal from a single welding pass.

welding current   The current flowing through a welding circuit during the making of a weld.

weldment   An assembly whose component parts are joined by welding.

weld metal   That portion of a weld that has been melted during welding.

Further Reading

Barer, R. D. and B. F. Peters, *Why Metals Fail*, Gordon and Breach Science Publishers, New York, 1970.

Case Histories in Failure Analysis, American Society for Metals, Metals Park, Ohio.

"Corrosion," *Metals Handbook*, edition 9, vol. 13, American Society for Metals, Metals Park, Ohio.

During, Evert D. D., *Corrosion Atlas* vol. 1, Elsevier Science Publishing Company, Inc., New York, 1988.

Evans, U. R., *The Corrosion and Oxidation of Metals, First Supplementary Volume*, Edward Arnold Ltd., London, 1968.

Evans, U. R., *Corrosion and Oxidation of Metals: Scientific Principles and Practical Applications*, Edward Arnold Ltd., London, 1960.

"Failure Analysis and Prevention," *Metals Handbook*, edition 8, vol. 10, American Society for Metals, Metals Park, Ohio.

Fontana, M. G. and N. D. Green, *Corrosion Engineering*, McGraw-Hill, Inc., New York, 1967.

French, D. N., *Metallurgical Failures in Fossil Fired Boilers*, John Wiley and Sons, Inc., New York, 1983.

Logan, Hugh L., *The Stress Corrosion of Metals*, National Bureau of Standards, Washington, D.C., John Wiley and Sons, Inc., New York, 1966.

Manual for Investigation and Correction of Boiler Tube Failures, Electric Power Research Institute, Palo Alto, California.

McCall, J. L. and P. M. French (eds.), *Metallography in Failure Analysis*, Plenum Press, New York, 1978.

McCoy, J. M., *The Chemical Treatment of Boiler Water*, Chemical Publishing Co., New York, 1981.

The Nalco Water Handbook, edition 2, McGraw-Hill Inc., New York, 1988.

Reid, William T., *External Corrosion and Deposits-Boilers and Gas Turbines*, American Elsevier Publishing Co., Inc., New York, 1971.

Shreir, L. L. (ed.), *Corrosion*, George Newness, Ltd., Tower House, London, 1963.

Speller, F. N., *Corrosion/Causes and Prevention*, McGraw-Hill, Inc., New York, 1951.

Steam/Its Generation and Use, edition 38, Babcock & Wilcox, New York, 1972.

Thielsch, Helmut, *Defects and Failures in Pressure Vessels and Piping*, Reinhold Publishing Corp., New York, 1965.

Uhlig, H. H. (ed.), *The Corrosion Handbook*, John Wiley and Sons, Inc., New York, 1948.

Specimen Index

Industry Index

Subject Index

About the Authors

ROBERT D. PORT is a metallurgical engineer with more than twenty years' experience in failure analysis, at both Nalco Chemical Company and at independent metallurgical laboratories. Mr. Port is the author of numerous papers on failure analysis, and frequently speaks to technical societies and industrial groups. He is a member of the National Association of Corrosion Engineers (NACE) and is active on its committee for Failure Analysis in Steam Generating Systems and its committee on Corrosion in Steam Generating Systems. He is currently president of the Nalco chapter of Sigma Xi, an international organization for research scientists. Mr. Port attended the University of Illinois where he received his B.S. in Metallurgical Engineering.

HARVEY M. HERRO obtained a B.S. in Physics from Marquette University and a Ph.D. in metallurgical engineering from Iowa State University. Dr. Herro, who has been with Nalco since 1982, holds patents for corrosion monitoring using chemical techniques, is the author of numerous papers on corrosion and failure analysis, and is a frequent lecturer. He is an active member in NACE and the American Society of Metals (ASM).

Mr. Port and Dr. Herro both have made presentations at professional societies such as the Electric Power Research Institute (EPRI), NACE, and ASM. Dr. Herro has also presented at the American Boiler Manufacturer's Association (ABMA). Together, the authors have completed more than 4,300 failure analyses.

Nalco Chemical Company, headquartered in Naperville, Illinois, has worldwide sales of over one billion dollars.